T0133118

INTELLIGENT SYSTEMS

Advances in Biometric Systems, Soft Computing, Image Processing, and Data Analytics

INTELLIGENT SYSTEMS

Advances in Biometric Systems, Soft Computing, Image Processing, and Data Analytics

Edited by

Chiranji Lal Chowdhary

Apple Academic Press Inc.
4164 Lakeshore Road
Burlington ON L7L 1A4
Canada

Apple Academic Press Inc.
1265 Goldenrod Circle NE
Palm Bay, Florida 32905
USA

Exclusive worldwide distribution by CRC Press, a member of Taylor & Francis Group

International Standard Book Number-13: 978-1-77188-800-4 (Hardcover)
International Standard Book Number-13: 978-0-42926-502-0 (eBook)

Library and Archives Canada Cataloguing in Publication

Title: Intelligent systems : advances in biometric systems, soft computing, image processing, and data analytics / edited by Chiranji Lal Chowdhary.

Other titles: Intelligent systems (Oakville, Ont.)

Names: Chowdhary, Chiranji Lal, 1975- editor.

Description: Includes bibliographical references and index.

Identifiers: Canadiana (print) 20190184701 | Canadiana (ebook) 20190184736 | ISBN 9781771888004 (hardcover) | ISBN 9780429265020 (ebook)

Subjects: LCSH: Intelligent control systems.

Classification: LCC TJ217.5 .I58 2020 | DDC 629.8—dc23

Library of Congress Cataloging-in-Publication Data

Names: Chowdhary, Chiranji Lal, 1975- editor.

Title: Intelligent systems : advances in biometric systems, soft computing, image processing, and data analytics / edited by Chiranji Lal Chowdhary.

Other titles: Intelligent systems (Apple Academic Press)

Description: First edition. | Oakville, ON ; Palm Bay, Florida : Apple Academic Press, [2020] | Includes bibliographical references and index. | Summary: "This volume, Intelligent systems : Advances in biometric systems, soft computing, image processing, and data analytics, helps to fill the gap between data analytics, image processing, and soft computing practices. Soft computing methods are used to focus on data analytics and image processing approaches to develop good intelligent systems. To this end, readers will find quality research that presents the current trends, advanced techniques, and hybridized techniques relating to data analytics and intelligence systems. The book also features case studies related to medical diagnosis with the use of image processing or soft computing algorithms in a particular models. Topics include some of the most important challenges and discoveries in intelligent systems today, such as computer vision concepts and image identification, data analysis and computational paradigms, deep learning techniques, face and speaker recognition systems, and more. Providing extensive coverage of biometric systems, soft computing, image processing, artificial intelligence, and data analytics, the chapter authors discuss the latest research issues, present solutions to research problems, and look at comparative analysis with earlier results. Key features present an overview on intelligent techniques available to develop intelligent systems. Provides informative coverage of biometric systems, soft computing, image processing, artificial intelligence, and data analytics. Focuses on the latest research issues, facilitates easy implementation of research to solve real-world problems. Presents novel algorithms and comparative result analysis"-- Provided by publisher.

Identifiers: LCCN 2019040900 (print) | LCCN 2019040901 (ebook) | ISBN 9781771888004 (hardcover) | ISBN 9780429265020 (ebook)

Subjects: LCSH: Systems engineering. | Artificial intelligence. | Expert systems (Computer science)

Classification: LCC TA168 .I485 2020 (print) | LCC TA168 (ebook) | DDC 006.3/3--dc23

LC record available at https://lccn.loc.gov/2019040900

LC ebook record available at https://lccn.loc.gov/2019040901

Apple Academic Press also publishes its books in a variety of electronic formats. Some content that appears in print may not be available in electronic format. For information about Apple Academic Press products, visit our website at **www.appleacademicpress.com** and the CRC Press website at **www.crcpress.com**

Printed and bound by CPI Group (UK) Ltd, Croydon, CR0 4YY

Dedicated to my beloved Parents,
Smt. Nainu Chowdhary and Shri Lumbha Ram Chowdhary.

—Chiranji Lal Chowdhary

About the Editor

Chiranji Lal Chowdhary

Chiranji Lal Chowdhary, PhD, is currently Associate Professor in the School of Information Technology & Engineering at VIT Vellore, India. He has 15 years of experience in academia and 6 months of experience in industry. He has received research awards for publishing research papers in refereed journals from VIT University fix times consecutively. He has guided more than 20 graduate projects and more than 20 postgraduate-level projects. His publications are indexed by the Clarivate Analytics, IEEE Computer Society, SCOPUS, the ACM Digital Library, and other abstract and citation databases. He is a reviewer for many reputed journals. He is a life member of the Computer Society of India, the Indian Science Congress Association, and the Indian Society for Technical Education. He has published many papers in refereed journals and has attended international conferences. He has contributed many chapters and is currently in the process of several editing books. His current research includes digital image processing, deep learning, pattern recognition, soft computing, and biometric systems.

Dr. Chowdhary received his PhD in information technology and engineering from VIT University; his MTech in computer science and engineering from (M. S. Ramaiah Institute of Technology, Bangalore), Visvesvaraya Technological University, Belagavi; and his BE in computer science and engineering from MBM Engineering College, Jai Narain Vyas University, India.

Contents

Contributors

K. Bhagyashree
Department of Software and System Engineering, School of Information Technology and Engineering, Vellore Institute of Technology, Vellore, India

Rachit Bhalla
Department of Software and System Engineering, School of Information Technology and Engineering, Vellore Institute of Technology, Vellore, India

Syed Muzamil Basha
Department of Information Technology, Sri Krishna College of Engineering and Technology, Coimbatore 641008, Tamil Nadu, India

D. K. Bebarta
Department of Computer Science and Engineering, GVPCEW, Vishakhapatnam, India

Chiranji Lal Chowdhary
Department of Software and System Engineering, School of Information Technology and Engineering (SITE), Vellore Institute of Technology, Vellore, 632014 Tamil Nadu, India

T. K. Das
School of Information Technology and Engineering, VIT University, Vellore, India

D. Dhanalakshmi
Department of Computer Science, Sri Ramakrishna College of Arts and Science, Coimbatore, India

Aman Goyal
Department of Software and System Engineering, School of Information Technology and Engineering, Vellore Institute of Technology, Vellore, India

Praveen Kumar Gupta
Department of Biotechnology, R. V. College of Engineering, Bangalore 560059, India

Lingayya Hiremath
Department of Biotechnology, R. V. College of Engineering, Bangalore 560059, India

Dharm Singh Jat
Department of Computer Science, Namibia University of Science and Technology, Namibia

Prashanth Kambli
Department of Information Science and Engineering, Ramaiah Institute of Technology, Bengaluru, India

S. Narendra Kumar
Department of Biotechnology, R. V. College of Engineering, Bangalore 560059, India

B. P. Vijay Kumar
Department of Information Science and Engineering, Ramaiah Institute of Technology, Bengaluru, India

Esha Kumar
Department of Software and System Engineering, School of Information Technology and Engineering, Vellore Institute of Technology, Vellore, India

Nivedhitha M.
School of Information Technology and Engineering (SITE), Vellore Institute of Technology, Vellore 632014, Tamil Nadu, India

Chucknorris Garikayi Madamombe
Department of Computer Science, Namibia University of Science and Technology, Namibia

Sanghamitra Mohanty
Department of Computer Science and Application, Utkal University, Bhubaneswar, India

E. Naresh
Department of Information Science and Engineering, Ramaiah Institute of Technology, Bengaluru, India

C. N. Patill
Department of Information Science and Engineering, Ramaiah Institute of Technology, Bengaluru, India

Dharmendra Singh Rajput
School of Information Technology and Engineering, Vellore Institute of Technology, Vellore, India

Sindhu Rajendran
Department of Electronics and Communication, R. V. College of Engineering, Bangalore 560059, India

M. Senthilkumar
School of Information Technology and Engineering, Vellore Institute of Technology, Vellore, India

Shubham Shrimali
Department of Software and System Engineering, School of Information Technology and Engineering, VIT Vellore, India

Gurpreet Singh
Department of Software and System Engineering, School of Information Technology and Engineering, Vellore Institute of Technology, Vellore, India

Gursimran Singh
Department of Software and System Engineering, School of Information Technology and Engineering, Vellore Institute of Technology, Vellore, India

Basanta Kumar Swain
Department of Computer Science and Engineering, Government College of Engineering, Kalahandi, Bhawanipatna, India

Meghamadhuri Vakil
Department of Electronics and Communication, R. V. College of Engineering, Bangalore 560059, India

Bhavesh Kumar Vasnani
Department of Software and System Engineering, School of Information Technology and Engineering, Vellore Institute of Technology, Vellore, India

Anna Saro Vijendran
School of Computing, Sri Ramakrishna College of Arts and Science, Coimbatore 641006, India

P. M. Durai Raj Vincent
School of Information Technology and Engineering (SITE), Vellore Institute of Technology, Vellore 632014, Tamil Nadu, India

Abbreviations

ΔMFCC	delta mel-frequency cepstrum coefficients
3DMFM	3D morphable face model
AAM	active appearance model
AI	artificial intelligence
CAD	computer-assisted diagnosis
CBIR	content-based image retrieval
CBR	case-based reasoning
CNNs	convolutional neural networks
CPN	counter propagation neural network
CRNN	cascade recurrent neural network
CT	computed tomography scans
DCT	discrete cosine transform
DFT	discrete Fourier transform
EBGM	elastic bunch graph matching
EHR	electronic health records
ES	expert system
FAR	false acceptance rate
FFT	fast Fourier transform
FMR	false match rate
FNMR	false non-match rates
FRR	false rejection rates
FTA	failure-to-acquire
FTE	failure to enroll
GFCC	gammatone frequency cepstral coefficients
GMM	Gaussian mixture model
GPUs	graphical preparing units
HE	histogram equalization
HMM	hidden Markov model
HOG	histogram of oriented gradients
HR	high resolution
ICA	independent component analysis
IR	infrared
LBP	local binary patterns

LDA	linear discriminant analysis
LEM	line edge map
LFA	local feature analysis
LPC	linear predictive coefficients
LPCC	linear predictive cepstral coefficients
LR	low resolution
LSTM	long-short-term memory networks
MAE	mean absolute error
MFCC	Mel-frequency cepstral coefficients
MMAE	maximum mean absolute error
MRI	magnetic resonance imaging
NLP	natural language processing
NPD	new product development
PET	positron emission tomography
PLPC	perceptual linear prediction coefficients
RBF	radial basis function
RDD	resilient distributed dataset
RM	risk minimization
RNN	recurrent neural network
ROI	region of interest
SC	soft computing
SIFT	scale-invariant feature transform
SMOTE	synthetic minority oversampling technique
SOM	self-sorting out guide
SR	super resolution
SVM	support vector machine
VGG	visual geometry group
VQ	vector quantization

Preface

Intelligent systems are available all around us and are being used to make our life easier. Such devices may be smart washing machines, digital televisions, automobiles, driverless vehicles, smart air traffic control systems, traffic lights, medical diagnosis systems, different biometric systems, classification and recognition systems, neural network systems, fuzzy systems, data analytic systems, and many more, which have great number of possibilities. In recent days, the Internet of things (IoT) is such an intelligent system that has almost everything imaginable, with unique identifiers and the capability to robotically transfer data over a network without requiring human-to-human or human-to-computer interaction (HCI). Researchers from assorted domains have shown a consistent urge to explore the rich contents in intelligent systems and techniques and to exploit the usefulness of new applications. Numerous methods have been proposed by the research, community to facilitate many techniques to develop intelligent systems and other existing or new algorithms elaborated with a detailed explanation. Intelligent systems are essential components for achieving sustainable development goals for our future generations to live with more facilities.

In the present information technology era, the existing systems are an unprejudiced assortment of associated essentials or mechanisms that are systematized for a common determination. The intelligent systems include not only intelligent devices but also interconnected assortments of such devices with the inclusion of networks and other superior systems. Also, intelligent systems can include refined artificial intelligence (AI)-based software systems, like expert systems, chatbots, and other types of software.

The aim of this book is to fill the gap between data analytics, image processing, and soft computing practices. Soft computing methods are used to focus on data analytics and image-processing approaches to develop good intelligent systems. To this end, readers will find quality research that presents the current trends, advanced techniques, and hybridized techniques relating to data analytics and intelligence systems. The book features case studies related to medical diagnosis with the use of image processing or soft computing algorithms in some particular models.

This volume provides extensive coverage of biometric systems, soft computing, image processing, AI, and data analytics. The objective is to discuss the latest research issues, to present solutions to research problems, and to look at comparative analysis with earlier results. One of the fundamental problems of computer vision is the description of any sensory data. For example, sensory data are texture, shape, color, motion, weight, smell, touch, sound, etc. Using computer vision concepts, we require an approach to descriptions of the type of objects that we may recognize as baby, lion, toy, pet, horse, or dog. However, 2D object-recognition techniques are sensitive to illumination, shadows, scale, pose, and occlusions. 3D object recognition, on the other hand, does not suffer from these limitations. 3D information of the scene from a particular point of view can be computed with the use of depth maps. These motivated us to move to 3D object recognition.

The identification process of an individual image (or an image from a video) by a human vision and a computer vision is for understanding the working of a 3D object-recognition system. A physical object, which is actually in 3D form, is captured by a sensor/camera (in case of computer vision) and seen by a human eye (in case of a human vision). When we are observing something, many other things are also involved, which makes it more challenging to recognize. After capturing such a thing by a camera or sensor, a digital image is formed that is nothing other than a bunch of pixels. It is becoming very important to know how a computer understands images. Various challenges are there in recognizing objects, such as viewpoint variations, illumination, occlusions, scale variations, deformation, background cluttering, and intra-class variations.

Soft computing has been an important research area where data analysis is concerned. It was developed as a significant analysis and computing tool to facilitate biological and linguistic inspiration for computational paradigms by highlighting fuzzy systems, artificial neural networks, rough sets, learning systems, genetic algorithms, and hybrid systems. For real-life problems and decision-making, researchers and experts require computational intelligence tools for appropriate planning and operation and for achieving proper results. Business intelligence basically deals with optimization and decision-making. Simultaneously, it must handle the uncertainties and impreciseness presentation of the data. Handing uncertainties and impreciseness in data analysis is not an easy task. Therefore, it is essential to think of some intelligent techniques in data analysis. Much research has been carried out in this direction where computer science is concerned. But the application of these computational intelligence techniques in business and management is rarely seen. The tools used in business intelligence today are more traditional than intelligent, and hence there is a need to implement these techniques in business intelligence and decision-making.

Chapter 1 presents an overview of intelligent techniques available to develop intelligent systems. Soft computing techniques such as expert systems, case-based reasoning, artificial neural networks, genetic algorithm, fuzzy systems are considered as artificial intelligence (AI) techniques since it involves some kind of human-like learning, decision-making, and acting. These techniques assist decision-makers to select effective actions in real-life especially in critical decision scenarios. The names of several soft computing tools are Fuzzy Systems, Neural Networks, Evolutionary Computation, Machine Learning, and Probabilistic Reasoning. Different soft computing tools can be used in different phases of the planned analytical models. Information processing and analysis such as removing noise, hierarchical classification and clustering, searching, decision-making, and predicting the data in order to build a smarter, efficient, adaptable system to assist human in decision-making in the fields of Information Systems & Business Intelligence, Internet computing, Image Processing, Robotics, Systems & Control, Bio-Engineering, and Financial Services & Engineering.

In this era, the biometric system plays a very important role in the various fields to protect and secure the important information or data of every individual. So that it verifies the authenticated user to access the information. In Chapter 2, some commonly used biometric system in real-time is fingerprint recognition, facial recognition, iris recognition, and palm vein recognition are discussed. Based on the compatibility these biometric systems are used in various applications such as are used in cell phones or mobiles, banking sector, government offices and so on. The artificial intelligence technology is used to provide the best feature or characteristic part of the individual for the identification and recognition purpose and to reduce the system complexity. Artificial Intelligence is used to provide efficient and proper identification of the individual using peculiar characteristics. Artificial intelligence techniques are used in the different biometric systems for efficient, security and privacy purposes. So that an only authenticated person will have to access their data. The different methodologies or algorithms of the artificial intelligence are used in the biometric system are explained in detail description.

Chapter 3, focuses on behavioral approach, that is, speech signal processing for the user authentication which seems to be a more complicated task in comparison to any physiological biometric technique. The research work is carried out on text dependent voice biometric system by hybridizing with a speech recognition engine for enhancing the performance of the biometric identification system. The key the reason behind the conglomeration of automatic speech recognition engine with the speaker recognition system is that it will fetch the information in physiological domain relating to the shape of the vocal tract (pharynx, larynx, mouth, jaw, tongue, and lungs) of an individual as well as the behavioral aspects, that is, place and manner of articulation.

In Chapter 4, the use of deep learning techniques predominantly the Convolutional Neural Networks (CNNs) has been used in various disciplines in recent years. CNNs have shown an essential ability to automatically extract large volumes of information from big data. The use of CNNs has significantly proved to be useful especially in classifying natural images. Nonetheless, there have been a major barrier in implementing CNNs in the medical domain due to lack of proper available data for training. As a result, generic imaging standards like ImageNet have been widely used in the medical domain. However, these generic imaging benchmarks are not so perfect as compared to the use of CNNs. The main aim of this chapter is review the existing deep learning techniques to classify and analyze medical imaging data. In this chapter, a comparative analysis of LeNet, AlexNet, and GoogLeNet was done. Furthermore, this review looked at the literature on classifying medicinal anatomy images using CNNs. Based on the various studies the CNNs architecture has better performance with the other architectures in classifying medical images. One of the most important traits in biometric systems is face recognition, which deals with the face images for identification, recognition, detection and classification of different people in individual or in group. Face recognition system is a biometric technology to identify or verify the user by comparing facial features to database. It is fast growing, challenging and interesting area in biometric systems.

It generally used for security purposes which can be compared with other biometrics like fingerprint, iris, etc. From the last 20 years, it becomes one of the most popular areas of research. As we know user privacy and the data are very important to protect it we need security. There are so many security systems or techniques to secure the data but every system has some loophole and to overcome this face recognition is used. In this system there is no need to remember the password, no one can steal your password. The password is your face and no one can access the data without your permission. In 1960, the first face recognition system was created. Since then, face recognition system comes a long way. There so many techniques came to improve the system and get more accurate result example Eigen face, Gabor filters, SVM, neural networks, 3D recognition, etc. Now with novel techniques and algorithms, it is more challenging to find out which is the best technique to use for security. Face recognition is one of the best security systems. These facial features are extracted like the size of nose, eyes, lips, etc. All these features are unique for every person. These features will be stored in the database whenever we want to verify the user we match these features to the database. There is some drawback like in case of twin's system can't differentiate which is a valid user. Second case if someone uses the photograph of user than the system can't differentiate that it is a real person or the photograph. In Chapter 5 a complete survey is done on the basis of possible techniques to overcome the drawbacks of security.

For communication, speech is one of the natural forms. A person's voice contains various parameters that convey information such as emotions, gender, attitude, health, and identity. Determination of these parameters will help to further develop the technology into a reliable and consistent means of identification using speaker recognition. Speaker recognition technologies have wide application areas especially in authentication; surveillance and forensic speaker recognition. The speaker recognition process based on a speech signal is treated as one of the most exciting technologies of human recognition. For Speaker identification activities we mainly emphasize the physical features of signal. Speakers could be categorized as speaker identification and speaker verification. In speaker identification, the obtained features are compared with all the speaker's features which are stored in a voice model database and in speaker verification the obtained features are only compared with the stored features of the speaker he/she claimed to be. In Chapter 6, the general principles of speaker recognition, methodology, and applications are discussed.

Chapter 7 has an analysis of the unimodal and multimodal biometric system. Biometric technology depends on the proposition of measuring and examining the biological traits of a person and extracting the particular features out of the collected data and then using it for comparisons with other templates stored in the database. These unique biological traits are often known as biometric identifiers and can be categorized into mainly two types. The first one being the physiological identifier and the second one being behavioral identifiers. The physiological identifiers are associated with the shapes and structure of the body, on the other hand, the behavioral identifiers are associated with the behavioral patterns of an individual. The physiological biometric identifiers consist of the following traits: Fingerprints, Face, Iris, Palm, Hand, Geometry, etc., whereas the behavioral consists of characteristics or behaviors such as voice, gait, typing rhythm, etc. In today's world, many different forms of access control have emerged which includes token-based identification systems, such as driver's license or passport, and knowledge-based identification systems such as a password or personal identification number. Even though these access controls are available biometric identifiers serve as a highly reliable and secure than the token and knowledge-based methods but there is one limitation the collection of biometric identifiers could increase the privacy concerns of the information's being used.

A novel heuristic approach of parameter tuning in SMOTE based preprocessing algorithm for imbalanced ordinal classification is proposed in Chapter 8. The main types of supervised learning problems include regression and classification problems. Ordinal classification resides in between classification and Regression. The ordinal classification shows its importance in the following fields. Facial recognition technology to classify the images based on places, time, etc., Recommendation engines suggest movies or television shows watch next based on user preference levels. The automobile engineering field needs the application of Ordinal classification in various decision-making processes. Class imbalance is the main cause to degrade the performance of the classifier. When a class imbalance exists in the ordinal multiclass scenario, it is very difficult for the classifier to learn from the existing available data. The nature of class imbalance distribution could occur in two situations such as when the class imbalance is an intrinsic problem or it happens naturally. When the data is not naturally imbalanced, instead it is too expensive to acquire such data for minority class learning due to cost, confidentiality and tremendous effort to find a well-represented dataset, like a very rare occurrence of the failure of a space shuttle. Class imbalance involves a number of difficulties in learning, including imbalanced class distribution, training sample size, class overlapping and small disjoints.

In Chapter 10, authors are proposing a novel method to combine the information from the visual and infrared image of the same scene using Counter Propagation Neural Network (CPN). Super-Resolution of Reconstruction of Infrared Images Adopting Counter Neural Networks. Tracking and recognition of targets based on infrared (IR) images has become an area of growing interest. Thermal infrared imagery provides a capability for identification under all lighting conditions including total darkness. Large information is provided by infrared images for higher temperature objects and less information for lower temperature objects. Visual image, on the other hand, provides the visual context to the objects. Coupling an infrared image with a visual image for additional spectral information and properly processing the two information streams has the potential to provide valuable information in the night and/or poor visibility conditions. Information provided by both sensors increases the performance of tracking and the robustness of the surveillance system. The image resolution is of importance to image processing. Higher image resolution holds more details, which is important to image recognition or image segmentation. Thus, it is desired to obtain a high-resolution (HR) image from its low-resolution (LR) counterpart(s). Due to the restrictions of the infrared capture device, infrared image resolution is generally low when compared with the visible image resolution. Super-resolution (SR) can generate an HR image from an LR image or a sequence of LR images. SR methods can be categorized into two classes, that is, multi-frame-based SR and learning-based SR. In the learning-based approach, the HDR image can be derived from its corresponding LR image with an image database.

In Chapter 11, high-end tools and technologies for managing data in the age of big data are discussed. Big Data is simply an enormous amount of data that is retrieved from heterogeneous sources (e.g., social media platforms, public, government, e-mail, E-learning platforms, etc.) and is capable of expanding over time. The data available can make a huge impact on the respective fields if it makes some sense, what is the use of having a meaningless data. To make some sense and gain insights for better prediction and decision-making from the boundless noisy data we need proper technologies and tools. An intelligent system is and embedded internet-connected machine which is capable of collecting and analyzing the data also communicate with many other systems. In other words, intelligent systems are machines that are advanced in technology which perceive and respond to the outside world but there are many challenges to developing an intelligent system with all the raw data available.

Chapter 12 makes an AI-based Chatbot using Deep Learning. A chatbot is brilliantly human-like machines that give conversation. The primary objective of the chatbot is to permit the clients and the machines to connect with each other to exchange their conversations. The way a machine can recognize human conversations and how they answer to the clients is challenging work. Chatbot utilizing sequence to sequence is developing an approach that provides a human-like conversation. This model builds a neural network that peruses a sentence and gives an extraordinary response. Chatbot using the neural network model gives extremely good outcomes in delivering a human verbal exchange. There are numerous present conversation model, however, they have some obstacles. Here, authors discuss sequence to sequence the neural network model, which ends up some responses by training with datasets. The sequence to sequence model have two primary modules encoder and decoder. The encoder typifies the target sentence to a middle representation. Then decoder uses that representation and creates a destination sentence. This model is applied for the open and close domain. The model can learn information from datasets. Utilizing domain-specific datasets can be used as solutions for certain issues.

In Chapter 13, authors are focusing on the use of deep learning concepts and analyzing real-time data based on location of users to detect effectively and rapidly crop diseases and pests. The images and other data like location will be collected using camera devices like mobile and sent to online server. The server will filter the possible diseases and pests by using previously collected data and downloaded data related to given location like temperature conditions. Then, neural networks can be used to figure out crop diseases/pests and result would be sent back to devices.

The image caption generation would be basically based on understanding long-short-term memory networks (LSTM) and recurrent neural networks. This network model allows us to select the next word of the sequence in a better manner. In image caption generation one platform offered is Python and with the help of the TensorFlow library, the user can easily generate the LSTM model for the given images. For this purpose, first, train the machine by a dataset for deep learning. To improve the efficiency of the caption generation, the training has to be quite deep with more sample images. In Chapter 14, deep learning based on recurrent neural networks (RNN) for image caption generation with greater accuracy is discussed. This chapter would be able to generate quite appropriate captions for the images that can be given as inputs. The model would concatenate all the fragments of a word to give an output. The applications of the proposed approach are quite wide, for example, caption generation for vloggers who post quite a lot of photos daily, for news channels to get real-time captions with a single pass. Later, there is a comparison and analyses the performance for this chapter.

Acknowledgments

It is my pleasure to express with deep sense of gratitude to Apple Academic Press for providing me this opportunity to edit this book, *Intelligent Systems: Advances in Biometric Systems, Soft Computing, Image Processing, and Data Analytics*. I would like to express my satisfaction and happiness that I have completed this project.

In a jubilant mood I express sincerely my whole-hearted thanks to all those who helped me in both direct and indirect ways to complete this work. First of all I would like to thank the authors who have contributed to this book. I acknowledge, with sincere gratitude, the kindness of the School of Information Technology and Engineering, Vellore Institute of Technology University, Vellore, India, for providing an opportunity to carry out this research work. In addition, I am also thankful to VIT for providing facilities to complete this project.

I take this unique opportunity to express my deepest appreciation to Sandra Jones Sickels, VP, Editorial and Marketing, Apple Academic Press Inc., for her effective suggestions, sincere instruction, and kind patience during this project. I am also thankful to Rakesh Kumar and Ashish Kumar, Apple Academic Press Inc., for their nice cooperation during this project.

I would like to thank our friends and faculty colleagues at VIT for the time they spared in helping us through difficult times. Special mention should be made of the timely help given by reviewers during this project, those whose names are not mentioned here. The suggestions they provided cannot be thrown into oblivion.

While writing, contributors have referenced several books and journals; I take this opportunity to thank all those authors and publishers. I am extremely thankful to the reviewers for their support during the process of evaluation.

Last but not least, I thank the production team of Apple Academic Press Inc. for encouraging me and extending their full cooperation and help in the timely completion of this book.

—Chiranji Lal Chowdhary

PART I

Biometric Systems and Image Processing

CHAPTER 1

Intelligent Techniques: An Overview

T. K. DAS[1*] and D. K. BEBARTA[2]

[1]*School of Information Technology and Engineering, VIT University, Vellore, India*

[2]*Department of Computer Science and Engineering, GVPCEW, Vishakhapatnam, India*

Corresponding author. E-mail: tapan.das@vit.ac.in

ABSTRACT

With the impression of enhancing the proficiency and effectiveness of image processing and biometric system, this article provides an insight to the leading intelligent techniques that could be employed to enrich the biometric image analysis process. Leading techniques like expert system, artificial neural network, fuzzy system, genetic algorithm, and rough computing are being investigated in this review for their suitability of implementing in a biometric system. However, one technique does not fit for all problem domains, hence basing on the complexity involved in underlying data; it has been found as which technique could be employed for a particular problem to elicit the desired knowledge out of it. This study can be helpful for designing an intelligent system, which can be tailored for the organizations basing on their domain and nature of data.

1.1 INTRODUCTION

Conventional computing known as hard computing needs an accurate analytical model with a lot of computation time. Real world problems exist in a non-ideal environment fails to compute accurately by many analytical models. Soft computing (SC) differs from hard computing and unlike hard computing, it is tolerant of imprecision, uncertainty, partial truth, and an approximation. The guiding principle of SC is to exploit the tolerance for imprecision, uncertainty, partial truth, and approximation to achieve tractability, robustness and low computation cost. The names of several SC tools are fuzzy systems, neural networks, evolutionary computation, machine learning, and probabilistic reasoning. Different SC tools can be used in different phases of the planned analytical models. Information processing and analysis such as removing noise, hierarchical classification and clustering, searching, decision-making, and predicting the data in order to build a smarter, efficient, adaptable system to assist human in decision-making in the fields of information systems and business intelligence, internet computing, image processing, robotics, systems and control, bioengineering, and financial services and engineering. Different SC techniques can be used to solve above-mentioned issues such as wavelets for removing noise, wavelets, fuzzy logic, and neural network for the hierarchical classification and clustering, evolutionary algorithms for searching, fuzzy systems for decision-making, artificial intelligence (AI) specifically suited to different tasks, such as waveform analysis, monitoring electronic data streams in the field of healthcare, energy market, currency exchange, stocks, and several other nonlinearity, unusual high volatility, and chaotic nature of data to predict the important trends.

SC techniques such as expert systems (ES), case-based reasoning (CBR), artificial neural networks, genetic algorithm (GA), fuzzy systems are considered as AI techniques since it involves some kind of human like learning, decision-making, and acting. These techniques assist decision makers to select effective actions in real-life especially in critical decision scenarios. Besides this, it reduces information overflow, facilitate current information; enable communication required for collaborative decisions; and deal with uncertainty in decision problems. Diverse range of intelligent techniques are represented in Figure 1.1.

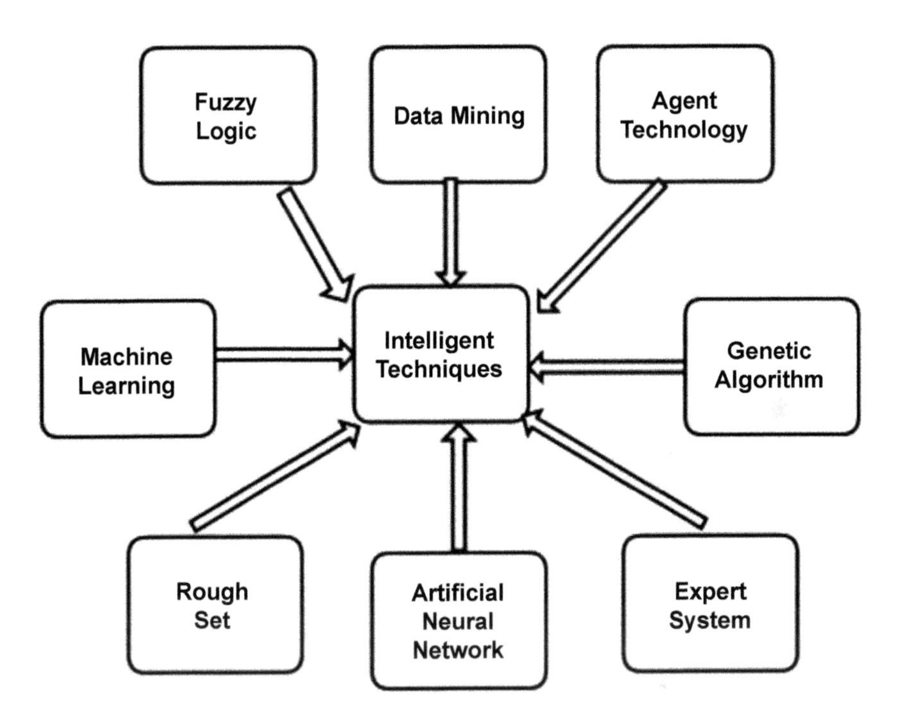

FIGURE 1.1 Intelligent techniques.
Source: Das (2016).

1.2 EXPERT SYSTEM

The inception of ES was long back in the year of 1972. The idea was to design a computer-based system which would work like a human expert. ES is a tool which augments human like decision-making by using AI techniques. It is programmed in such a way that it could be able to achieve the human logical thinking ability. It is a computer-based system programmed to use AI technique. Initially knowledge is acquired from domain experts and the acquired knowledge is represented in various forms such as rules, frames, or semantic nets. The heart of ES is inference engine which generates inference out of knowledgebase. An abstract view of ES is represented in Figure 1.2.

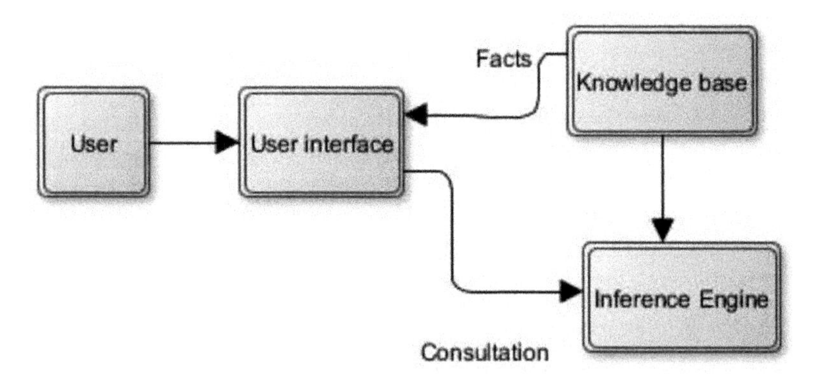

FIGURE 1.2 Abstract view of an expert system.

ES is being built for a variety of purposes, including medical diagnosis, mineral exploration, system fault finding, and many more. It is used in a context when a person requires some expertise to solve a problem. It is beneficial when human experts are expensive and difficult to find. However, it is quite useful in a situation of unpleasant environment and monotonous operation. One of the first generation of ES meant for chemical analysis; DENDRAL (Liang and Mahmud, 2012) Healthcare domain ES monitors surgeries, diagnosis the problem, and suggest actions. ES also automatically collect, aggregate, and analyze data for detailed analysis. ES, for example, MYCIN (Shortliffe, 1976), PROSPECTOR (Duda and Reboh, 1984) can handle uncertainty aspect of the underlying data as well.

CBR is quite effective in analyzing and designing decision support system for diagnosis in healthcare domain (Bichindaritz and Marling, 2006). CBR systems have been designed to look at the complexity of biomedicine, to directly integrate into clinical settings for closely monitoring and communicating with diverse systems. CBR system usefulness in decision-making in integrating with web 2.0 has been studied by He et al. (2009).

1.3 ARTIFICIAL NEURAL NETWORK

Artificial neural network known as ANN comprises a collection of simple nonlinear (or linear) processing components called neurons, which are linked together via a net of adaptable connections. A neuron can transform a set of input to a reasonable output depending on the properties of activation function it uses. A set of neurons act as input nodes and another set act as output nodes. In between input and output layers, the nodes process information by weighted summation, and thresholding. These operations are guided by the principle of a leaning algorithm basing on which the network functions. Learning for a network is known as training. Like human being, the network is being trained in order to accomplish a particular task. A multilayer neural network is shown in Figure 1.3.

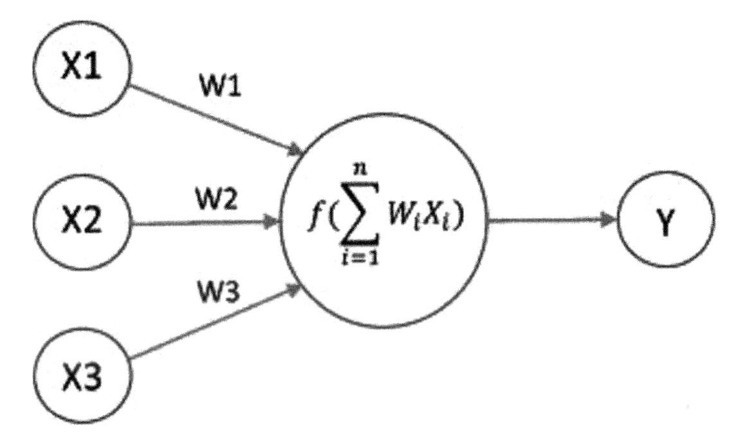

FIGURE 1.3 The simple view of a neural network.

The neural network develops solutions that are far beyond the ambit of the learning data; which makes usefulness of ANN in several applications (Pedrycz et al., 2008; Das, 2015). Likely, some other Chowdhary and Achrjya (2018), Chowdhary (2011), Chowdhary et al. (2015), (2016).

There is a plethora of instances where neural networks has been applied for decision-making (Saridakis and Dentsoras, 2006; Chen and Lin, 2003; Gholamian et al., 2006). Besides, complex mathematical problems are clearly symbolized using neural network presentations (Li and Li, 1999). Furthermore, it has been perceived that the success of ANNs in the financial domain is quite promising, especially in financial prediction and planning, financial evaluation, credit approval and determining optimal capital structure (Bahrammirzaee et al., 2010; Bebarta et al., 2012). Hence it has become an effective decision-making tool in financial domain (Lam, 2004) especially in stock market prediction (Bebarta et al., 2015). Table 1.1 lists other application areas of ANN.

TABLE 1.1 Applications of Neural Network.

Application field	Sub-field	Examples	References
Image processing	Feature extraction, prediction, pattern recognition, and representation/transformation	Neural filters for removing quantum noise, neural edge detector, neural edge enhancer, reduction of false positives	Suzuki et al. (2002), (2003)
Robotics (systems and control)	Primary visual cortex, posterior parietal cortex (PPC)	Motion energy model, steer collision-free paths	Beyeler et al. (2015) and Sun and Joo (2004)
Information systems and business intelligence	Sentiment analysis, text analytics, web mining	E-commerce and market analytics, e-Government	Chen et al. (2012)
Petroleum and natural gas engineering	Oil and gas	Pressure-transient analysis, well-log interpretation, reservoir characterization, candidate-well selection for stimulation,	Mohaghegh (2000)
Biomedical engineering/ healthcare	Clinical diagnosis, image and signal analysis	Fall detection and activity monitoring for oldsters, body motion information collection, remote monitoring and analyzing patients, heartbeat classification using feature selection, MRI, CT scan, ECG, and so on.	Yardimci (2009)

1.4 FUZZY SYSTEM

This theory was introduced in 1965 by L. Zadeh. In classical set theory, objects are considered either as being a member of a set or not a member (black or white). Hence it is also referred as crisp set. Therefore, there are only two values for the membership: 1 (member) and 0 (non-member); there is no scope for any value apart from 0 and 1.

On the other hand, fuzzy set claims that membership is concerned about degree of belongingness and it lies within the range of 0 and 1. Fuzzy logic is capable of portraying, to a reasonable extent, human type reasoning in natural form. Modeling of imprecise and qualitative knowledge, as well as handling of uncertainty is possible through a fuzzy system (Pedrycz, 1998). The working of a fuzzy system is represented in Figure 1.4. Fuzzy set have been used in many application domains to quantify the uncertainty for decision-making tasks (Iliadis, 2005). In addition to this, fuzzy sets are found to be effective in handling multi-objective decision-making processes. Fuzzy-logic-based decision-making approaches are being applied for risk assessment in the course of new product development (NPD) phase to determine the appropriate decision models and techniques (Wang, 1997; Buyukozkan and Feyzıoglu, 2004). Besides, fuzzy logic is also used in exploring knowledge from a database (Yager, 1996).

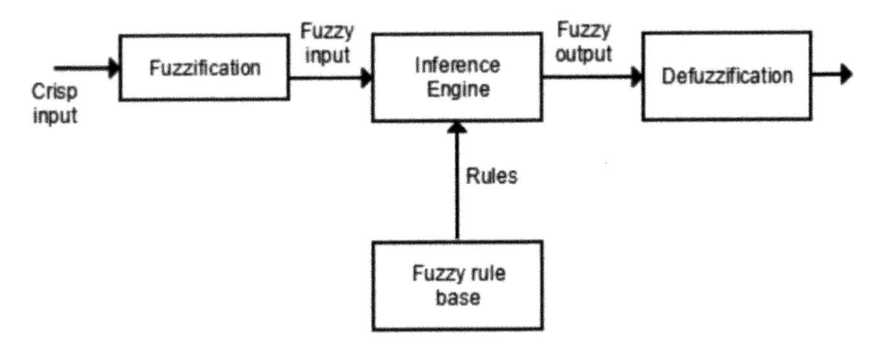

FIGURE 1.4 Architecture of a fuzzy controller.

1.5 GENETIC ALGORITHM

Most of the scientific and engineering problems are solved using trial and errors methods using various optimization techniques. Many researchers have introduced various optimization techniques; GA, Differential Evolution algorithm, Particle Swarm Optimization, Bacterial Foraging, Clonal algorithm, Firefly algorithm, Memetic algorithm, Ant Colony Optimization, Frog Leaping algorithm, and many more.

GA is a search algorithm based on the characteristics of evolutionary approach (Goldberg and Holland, 1988). It is a population-based evolution mechanisms of reproduction, crossover, and mutation to provide a robust solution to complex search space. GA is an iterative computation process, and its main steps include: encoding, initialization of the population, selection, genetic operation (crossover, mutation), evaluation, and stop decision (Yao and Xu, 2006).

The detailed operations of GA are represented in Figure 1.5.

Evolutionary computing has been applied in decision-making tasks in several ways (Pedrycz et al., 2008); few of them are listed below:

1. Due to possible structures of the models, structural optimization becomes crucial in many cases. In this juncture, evolutionary optimization helps for choosing an optimal one. Hence, GA as it is a manifestation of evolutionary optimization is quite efficiently been used in this case. (Jiang et al., 2009).
2. Decision-making processes typically encounter a number of criteria to judge; these particular difficulties known as multicriteria nature call for their simultaneous optimization and evolutionary techniques are favorable to address this kind of problems (Fonseca and Fleming, 1998).
3. However, it is worthwhile to study a collection of possible solutions rather than the optimal one; including those that are suboptimal; these could offer a better insight about the nature of the decision-making problem and allow for a global characterization of the solutions (Jakob et al., 1992).
4. Genetic programing involving genetic operators (Koza, 1994; Kinnear, 1994) is introduced in order to translate it into computer programs.

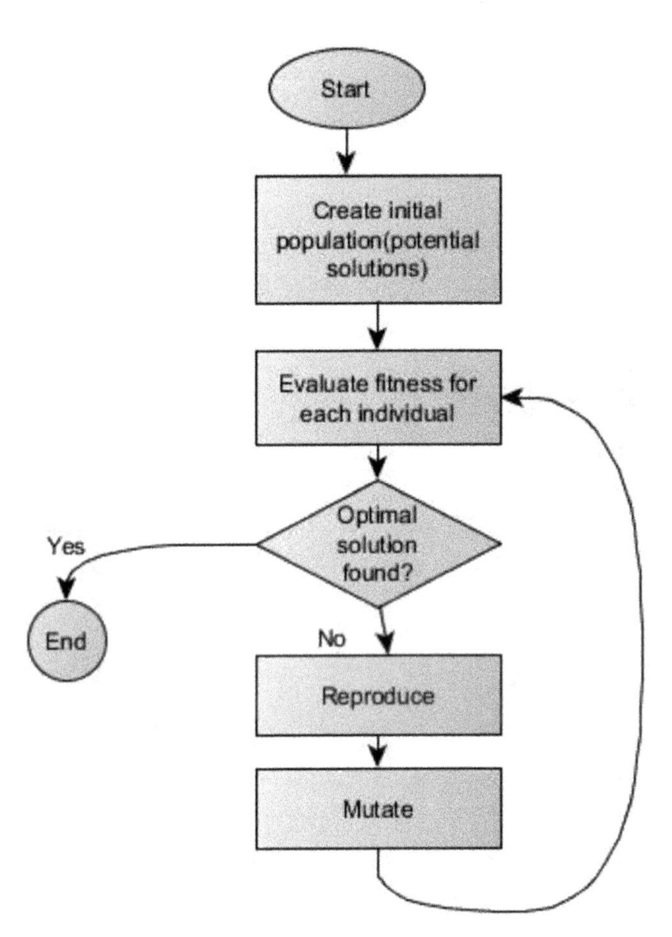

FIGURE 1.5 Genetic algorithm process.

1.6 ROUGH SET

Rough set theory was introduced by Z. Pawlak (Pawlak, 1982) in the year 1982. Unlike fuzzy set which expresses vagueness by means of membership, it creates a boundary region for the target set. The elements who do fall on the boundary line neither be classified in either category. This uncertainty is approximated as lower and upper approximation, that is, a set of elements which are definitely belongs and another set which possibly belongs in target set.

Rough set theory is found to be a leading intelligent technique as it has proved its efficacy almost in all areas (Chowdhary and Acharjya, 2016, 2017; Das and Chowdhary, 2017). A few of them are listed below:

1. Rule extraction from an information system which generally has few attributes and a decision class attribute (Hu and Cercone, 1996; Skowron, 1995). Few applications where decision rules are extracted from clinical datasets (Tsumoto, 2003) and online reviews (Das et al., 2014) generates of discernibility matrices and reducts in between.
2. Rough set is a step ahead as compared to other theory in selecting features which also synonymously known as attribute reduction (Kira and Rendell, 1992; John et al., 1994; Swiniarski and Skowron, 2003; Thangavel and Pethalakshmi, 2009; Acharjya and Das, 2017). Attribute reduction is about choosing a subset of original attributes by eliminating the irrelevant ones in order to improve the performance of the subsequent processing of the data.
3. Processing incomplete information system by means of attribute reduction followed by classification (Grzymala-Busse, 2003; Thangavel et al., 2005).
4. Intelligent decision-making tasks, for example, for patients with lung abnormalities (Kusiak et al., 2000). Further rough set is being employed for various decision-making tasks in almost all domains (Das and Acharjya, 2014; Das et al., 2015).

1.7 CONCLUSION

Intelligent systems work on the guiding principles of intelligent techniques and associated algorithms perform complex decision-making tasks. Many real-world problems are solved with collaborative techniques known as hybrid models; by combining artificial neural network with fuzzy logic, GAs with fuzzy logic, and GA with an artificial neural network. However, the success of intelligent systems is attributed to greater processing power of computer and ample storage space in a reduced cost. Even intelligent techniques can be programmed by a low cost microcontroller. In the era of AI and automation, intelligent technique concepts are being employed even in consumer electronics, household appliances. Hence it has become an integral part in the domain applications starting from home science till rocket science.

KEYWORDS

- **fuzzy systems**
- **genetic system**
- **hard computing**
- **intelligent systems**
- **rough set**
- **soft computing**

REFERENCES

1. Acharjya, D. P.; Das, T. K. A Framework for Attribute Selection in Marketing Using Rough Computing and Formal Concept Analysis. *IIMB Manag. Rev.* **2017,** *29*, 122–135.
2. Bahrammirzaee, A. A Comparative Survey of Artificial Intelligence Applications in Finance: Artificial Neural Networks, Expert System and Hybrid Intelligent Systems. *Neural Comput. Appl.* **2010,** *19*, 1165–1195.
3. Beyeler, M.; Oros, N.; Dutt, N.; Krichmar, J. L. A GPU-Accelerated Cortical Neural Network Model for Visually Guided Robot Navigation. *Neural Netw.* **2015,** *72*, 75–87.
4. Bichindaritz, I.; Marling, C. Case-based Reasoning in the Health Sciences: What's Next? *Artif. Intell. Med.* **2006,** *36*, 127–135.
5. Buyukozkan, G.; Feyzıoglu, O. A Fuzzy-logic-based Decision-making Approach for New Product Development. *Int. J. Prod. Econ.* **2004,** *90*, 27–45.
6. Chen, H.; Chiang, R. H. L.; Storey, V. C. Business Intelligence and Analytics: From Big Data to Big Impact, Special Issue on Business Intelligence Research. *MIS Q.* **2012,** *36* (4), 1165–1188.
7. Chen, J.; Lin, S. An Interactive Neural Network-based Approach for Solving Multiple Criteria Decision-making Problems. *Dec. Supp. Syst.* **2003,** *36*, 137–146.
8. Chowdhary, C. L. Application of Object Recognition With Shape-Index Identification and 2D Scale Invariant Feature Transform for Key-Point Detection. Feature Dimension Reduction for Content-Based Image Identification, 2018, pp 218–231.
9. Chowdhary, C. L.; Acharjya, D. P. A Hybrid Scheme for Breast Cancer Detection Using Intuitionistic Fuzzy Rough Set Technique. *Biometr. Concepts Method. Tools Applic.* **2016,** 1195–1219.
10. Chowdhary, C. L.; Acharjya, D. P. Clustering Algorithm in Possibilistic Exponential Fuzzy c-mean Segmenting Medical Images. *J. Biomimet. Biomater. Biomed. Eng.* **2017,** *30*, 12–23.

11. Chowdhary, C. L.; Acharjya, D. P. Singular Value Decomposition–Principal Component Analysis-Based Object Recognition Approach. *Bio-Insp. Comput. Image Video Process.* **2018,** 323.
12. Chowdhary, C. L.; Muatjitjeja, K.; Jat, D. S. Three-dimensional Object Recognition Based Intelligence System for Identification. *Emerg. Trends Netw. Comput. Commun.* **2015**.
13. Chowdhary, C. L.; Ranjan, A.; Jat, D. S. Categorical Database Information-Theoretic Approach of Outlier Detection Model. *Ann. Comput. Sci. Ser.* 14th Tome 2nd Fasc. **2016,** *2016*, 29–36.
14. Das, T. K.; Acharjya, D. P.; Patra, M. R. In *Business Intelligence From Online Product Review—A Rough Set Based Rule Induction Approach*, 2014 International Conference on Contemporary computing and informatics (IC3I- 2014), IEEE, 800-803, Mysore, India.
15. Das, T. K. In *A Customer Classification Prediction Model Based on Machine Learning Techniques*. International Conference on Applied and Theoretical Computing and Communication Technology (iCATccT), 2015; pp 321–326.
16. Das, T. K. Intelligent Techniques in Decision Making: A Survey. *Ind. J. Sci. Technol.* **2016,** *9* (12), 1–6.
17. Das, T. K.; Acharjya, D. P. A Decision Making Model Using Soft Set and Rough Set on Fuzzy Approximation Spaces. *Int. J. Intell. Syst. Technol. Applic.* **2014,** *13* (3), 170–186.
18. Das, T. K.; Chowdhary, C. L. Implementation of Morphological Image Processing Algorithm Using Mammograms. *J. Chem. Pharm. Sci.* **2016,** *10* (1), 439–441.
19. Das, T. K.; Acharjya, D. P.; Patra, M. R. General Characterization of Classifications in Rough Set on Two Universal Sets. *Inf. Resour. Manag. J.* **2015,** *28* (2), 1–19.
20. Duda, R. O.; Reboh, R. *AI and Decision Making: The PROSPECTOR Experience. Artificial Intelligence Applications for Business*; Ablex Publishing Corp.: Norwood, NJ, 1984.
21. Fonseca, C. M.; Fleming, P. J. Multiobjective Optimization and Multiple Constraint Handling with Evolutionary Algorithms–Part I: A Unified Formulation. *IEEE Trans. Syst. Man Cybern. Part A Syst. Humans* **1998,** *28* (1), 26–37.
22. Gholamian, M. R.; Fatemi, S. M. T. G.; Ghazanfari, M. A Hybrid Intelligent System for Multi-Objective Decision Making Problems. *Comput. Ind. Eng.* **2006,** *51*, 26–43.
23. Goldberg, D. E.; Holland, J. H. Genetic Algorithms and Machine Learning. *Mach. Learn.* **1988,** *3* (3), 95–99.
24. Grzymala-Busse, J. W. (2003). In *Rough Set Strategies to Data with Missing Attribute Values*. Proceedings of the Workshop on Foundations and New Directions in Data Mining, Associated with the 3rd IEEE International Conference on Data Mining, 2003; pp 56–63.
25. He, W.; Xu, L. D.; Means, T.; Wang, P. Integrating Web 2.0 With the Case-Based Reasoning Cycle: A Systems Approach. *Syst. Res. Behav. Sci.* **2009,** *26* (6), 717–728.
26. Hu, X.; Cercone, N. In *Mining Knowledge Rules From Databases: A Rough Set Approach*. Proceedings of the 12th International Conference on Data Engineering; Washington, DC, 1996; pp 96–105.
27. Iliadis, L. S. A Decision Support System Applying an Integrated Fuzzy Model for Longterm Forest Fire Risk Estimation. *Environ. Model. Softw.* **2005,** *20*, 613–621.

28. Jakob, W. D.; Schleuter, M. G.; Blume, C. Application of Genetic Algorithms to Task Planning and Learning. In *Parallel Problem Solving From Nature*, M¨anner, R., Manderick, B., Eds.;North-Holland: Amsterdam, The Netherlands, 1992; Vol. 2, pp 291–300.
29. Jiang, Y.; Xu, L.; Wang, H. Influencing Factors for Predicting Financial Performance Based on Genetic Algorithms. *Syst. Res. Behav. Sci.* **2009,** *26* (6), 661–673.
30. John, G. H.; Kohavi, R.; Pfleger, K. In *Irrelevant Features and the Subset Selection Problem*. Proceedings of 11th International Conference on Machine Learning, 1994; pp 121–129.
31. Kinnear, K. E. *Advances in Genetic Programming*; The MIT Press: USA, 1994.
32. Kira, K.; Rendell, L. A. *The Feature Selection Problem: Traditional Methods and a New Algorithm*, Proceedings of AAAI; MIT Press, 1992; pp 129–134.
33. Koza, J. R. *Genetic Programming 2*; MIT Press: Cambridge, MA, 1994.
34. Kusiak, A.; Kern, J. A.; Kernstine, K. H.; Tseng, B. T. L. *Autonomous Decision-Making: A Data Mining Approach. IEEE Trans. Inf. Technol. Biomed.* **2000,** *4* (4), 274–284.
35. Lam, M. Neural Network Techniques for Financial Performance Prediction, Integrating Fundamental and Technical Analysis. *Dec. Sup. Syst.* **2004,** *37*, 567–581.
36. Li, H.; Li, L. Representing Diverse Mathematical Problems Using Neural Networks in Hybrid Intelligent Systems. *Exp. Syst.* **1999,** *16* (4), 262–272.
37. Liang, Y. W.; Mahmud, R. In *A Comparison Model for Uncertain Information in Expert System*. IEEE International Conference on Uncertainty Reasoning and Knowledge Engineering, 2012; pp 127–130.
38. Mohaghegh, S. *Virtual-Intelligence Applications in Petroleum Engineering: Part 1, Artificial Neural Networks*, SPE Distinguished Author Series, 2000.
39. Pawlak, Z. Rough Sets. *Int. J. Comput. Inf. Sci.* **1982,** *11*, 341–356.
40. Pedrycz, F. G. *An Introduction to Fuzzy Sets: Analysis and Design*; MIT Press: Cambridge, MA, 1998.
41. Pedrycz, W.; Ichalkaranje, N.; Phillips-Wren, G.; Jain, L. C. Introduction to Computational Intelligence for Decision Making. In *Intelligent Decision Making: An AI-Based Approach*, Phillips-Wren, G., Ichalkaranje, N., Jain, L. C., Eds.; Springer-Verlag: Berlin, Heidelberg, 2008; Vol. 97, pp 79–97.
42. Saridakis, K. M.; Dentsoras, A. J. Integration of Fuzzy Logic, Genetic Algorithms and Neural Networks in Collaborative Parametric Design. *Adv. Eng. Inform.* **2006,** *20*, 379–399.
43. Shortliffe, E. H. *Computer-based Medical Consultations: MYCIN,* 388; Elsevier: New York, 1976.
44. Skowron, A. Extracting Laws From Decision Tables—A Rough Set Approach. *Comput. Intell.* **1995,** *11*, 371–388.
45. Sun, Y. L.; Joo, M. Hybrid Fuzzy Control of Robotics Systems. *IEEE Trans. Fuzzy Syst.* **2004,** *12* (6), 755–765.
46. Suzuki, K.; Horiba, I.; Sugie, N. Neural Edge Enhancer for Supervised Edge Enhancement From Noisy Images. *IEEE Trans. Pattern Anal. Mach. Intell.* **2003,** *25* (12), 1582–1596.
47. Suzuki, K.; Horiba, I.; Sugie, N. Efficient Approximation of Neural Filters for Removing Quantum Noise From Images. *IEEE Trans. Signal Process.* **2002,** *50* (7), 1787–1799.

48. Swiniarski, W.; Skowron, A. Rough Set Methods in Feature Selection and Recognition. *Pattern Recogn. Lett.* **2003,** *24* (6), 833–849.
49. Thangavel, K.; Pethalakshmi, A. Dimensionality Reduction Based on Rough Set Theory: A Review. *Appl. Soft Comput.* **2009,** *9,* 1–12.
50. Thangavel, K.; Jaganathan, P.; Pethalakshmi, A.; Karnan, M. Effective Classification With Improved Quick Reduct for Medical Database Using Rough System. *Bioinform. Med. Eng.* **2005,** *5* (1), 7–14.
51. Tsumoto, S. Automated Extraction of Hierarchical Decision Rules From Clinical Databases Using Rough Set Model. *Exp. Syst. Appl.* **2003,** *24* (2), 189–197.
52. Wang, J. A Fuzzy Outranking Method for Conceptual Design Evaluation. *Int. J. Prod. Res.* **1997,** *35* (4), 995–1010.
53. Yager, R. R. Database Discovery Using Fuzzy Sets. *Int. J. Intel. Syst.* **1996,** *11,* 691–712.
54. Yao, X.; Xu, Y. Recent Advances in Evolutionary Computation. *J. Comput. Sci. Technol.* **2006,** *21* (1), 1–18.
55. Yardimci, A. Soft Computing in Medicine. *Appl. Soft Comput.* **2009,** *9* (3), 1029–1043.

CHAPTER 2

A Survey of Artificial Intelligence Techniques Used in Biometric Systems

C. N. PATILL*, E. NARESH, B. P. VIJAY KUMAR,
and PRASHANTH KAMBLI

*Department of Information Science and Engineering,
Ramaiah Institute of Technology, Bengaluru, India*

*Corresponding author. E-mail: nareshkumar.e@gmail.com

ABSTRACT

In this era, the biometric system plays a very important role in the various fields to protect and secure the important information or data of every individual. So that it verifies the authenticated user to access the information. There are some commonly used biometric system in real-time that are Fingerprint recognition, Facial recognition, Iris recognition, and Palm vein recognition. Based on the compatibility these biometric systems are used in various applications such as are used in a cell phone or mobiles, banking sector, government offices, and so on. Artificial intelligence technique is used to provide the best feature or characteristic part of the individual for the identification and recognition purpose and to reduce the system complexity. Artificial intelligence is used to provide efficient and proper identification of the individual using peculiar characteristics. Artificial intelligence techniques are used in the different biometric system for efficient, security, and privacy purposes. So that an only authenticated person will have access to his/her data. The different methodology or algorithm of artificial intelligence are used in the biometric system are explained in detail description.

2.1 INTRODUCTION

Nowadays biometrics is one of the most trending technologies in various fields, used for security and privacy purpose. Biometrics is used for the recognition and authentication purpose of an individual and for the security and privacy purpose to secure the data access by the unauthorized person or entity. Fingerprints, facial recognition, iris biometrics, and retina biometrics are some of the physical biometric identification, recognition, and authentication methods. Because of the unique signature of an individual, the unauthorized person or entity cannot access the device or other personal information of an individual. Artificial intelligence technique is used to provide the best feature or characteristic part of the individual for the identification and recognition purpose and to reduce the system complexity. Biometric system works with peculiar characteristics feature part of the human body. Because of the high quality of clarity of image, it acquires a very large amount of space and then gradually it reduces the efficiency of the biometric system. So, artificial intelligence is used to provide efficient and proper identification of the individual using peculiar characteristics. Human face recognition plays a very important role in biometric applications that are used for photography, human–computer interaction, artificial intelligence, to unlock the mobile devices and in various security applications.[19–24]

In this era, the biometric system plays a very important role in the various fields to protect and secure the important information or data of every individual. So that it verifies the authenticated user to access the information. There are some commonly used biometric systems in real time, namely, fingerprint recognition, facial recognition, iris recognition, and palm vein recognition. Based on the compatibility these biometric systems are used in various applications, such as in a cell phone or mobiles, banking sector, government offices, and so on. Artificial intelligence technique is used to provide best feature or characteristic part of the individual for the identification and recognition purpose and to reduce the system complexity. Artificial intelligence is used to provide efficient and proper identification of the individual using peculiar characteristics. Artificial intelligence techniques are used in different biometric systems for efficient security and privacy purposes. So that the only authenticated person will have access to their data. Different methodology or algorithms of artificial intelligence are used in biometric systems that are explained in detail description.

This chapter is organized as follows: Section 2.2 illustrates characteristic of various biometric systems, Section 2.3 emphasizes on facial recognition, Section 2.4 is covering finger recognition system, Section 2.5, iris recognition, and the last section says about the conclusion and future work.

2.2 CHARACTERISTICS OF VARIOUS BIOMETRIC SYSTEMS

Biometric-system identification and recognition method is based on the characteristics and unique identity of every individual. The characteristics and unique identity of every individual is based on physical and behavioral traits. In the biometric system there are some key advantages, like they are nontransferable, nonrepudiation, nonassumable, and gives more protection against the fraud identification and recognition. These technologies are successfully implemented in various real-time applications, such as the banking sector, financial institutions, government offices, and educational institutes, company identity management, and other identification and recognition purposes. There are some most commonly used biometric systems for the identification and recognition method, including facial recognition, palm vein recognition, iris recognition, fingerprint recognition, and voice recognition.

2.2.1 FACIAL RECOGNITION

Facial recognition is one of the most recently developed identification and recognition method in the biometric system. Facial recognition provides a prominent and efficient identification of an individual even in the masses or group of people within a fraction of second using a front scanner, that is, digital camera (Fig. 2.1). Facial recognition system (Fig. 2.2) analyzes some certain key features that are common in each and every individual face. Every individual can be identified by some of the features that include the distance between the two eyes, shape and length of nose, shape of mouth, chin, cheekbones, and position of cheekbones, etc., by combining those features or numerical quantities into a single code. In face recognition illumination condition, pose, color, and brightness are major challenging problems.[1,2] AdaBoost approach and skin-based segmentation is most commonly used techniques for the face recognition and authentication. But skin color-based segmentation technique provides a high rate of false detection of image with a complicated background, and AdaBoost algorithm is unable to detect the images having multiple faces and multiple poses. AdaBoost approach is used for the high accuracy and skin-based segmentation method used for the fast detection of the image. To improve the performance of both detection and accuracy morphological operators are applied. As a result, the face recognition calculation enhances the discovery of speed exactness and enables to do ongoing identification of image with high accuracy, and it can detect the faces of various size, position, and expression. Figure 2.1 represents the face image along with background.[14,15] Various object recognition approaches can also be used to improve biometric systems.[17–23]

FIGURE 2.1 Face image with background.

Source: Ref. [15].

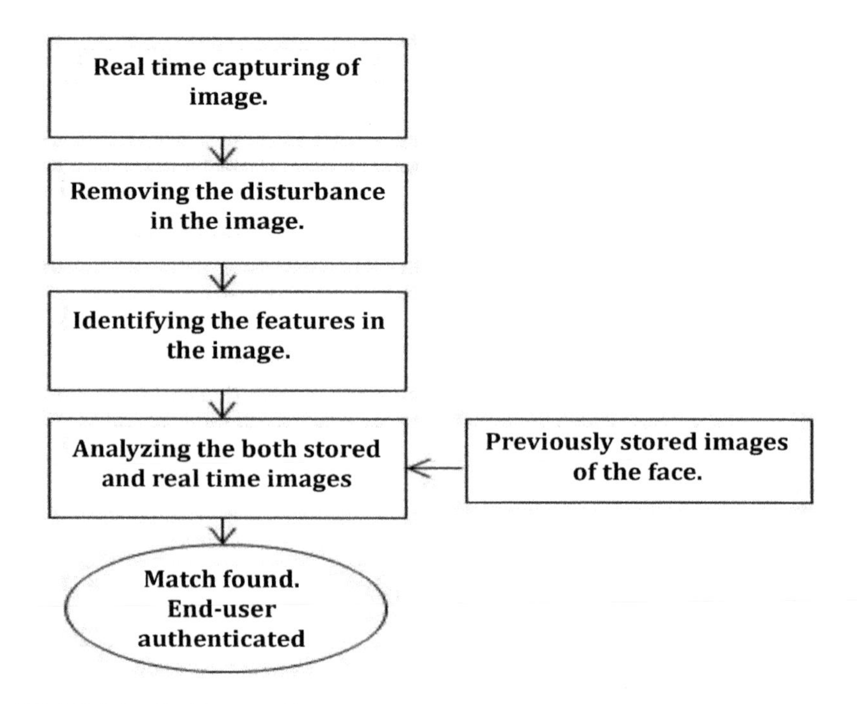

FIGURE 2.2 Facial recognition process.

2.2.1.1 AdaBoost ALGORITHM

In recent technology, development in the areas of security and privacy facial identification and recognition or facial detection biometric system plays an important role. The AdaBoost algorithm is one of the best and quick approaches for face detection. There are mainly three principles that are followed by the AdaBoost algorithm approach.

2.2.1.1.1 Computation of Rectangular Feature

In facial recognition biometric system, face image of the user is captured, and Haar-like feature is used to recognize the object in the real-time captured image. Haar-based cascade classifiers are used to distinguish between various types of Haar features, like Edge feature, Line feature, four rectangle feature, and center-surrounded features. Haar features are provided as a contribution to the course classifiers (Fig. 2.3). This Haar feature can be calculated by assuming the contrast between aggregate of pixels in white rectangular part or section and aggregate of pixels in dark rectangular area at various aspects of orientations, proportions, and scales.[8]

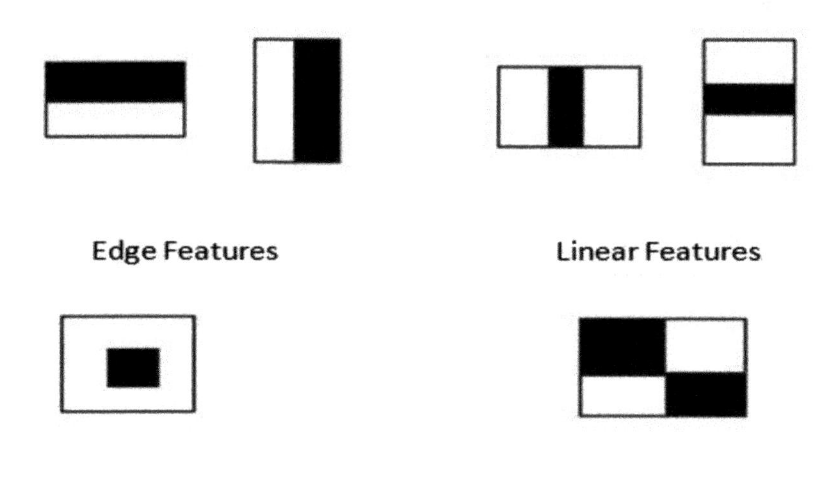

FIGURE 2.3 Different types of image features.

2.2.1.1.2 Improved AdaBoost Algorithm

Using the AdaBoost algorithm, several weak classifiers are combined into one strong classifier. Improved AdaBoost algorithm is applied to training sample to allocate with a weight, so it can be easy to choose the preparation set by some grouping based on the percentage of the weight it represents.[11,12,14] The AdaBoost algorithm is explained below:

1. Let given sample images $(X_1, Y_1) \dots \dots (X_L, Y_L)$, where $Y_i = 0$ represents the negative image and $Y_i = 1$ represents the positive image, respectively.
2. Initialization of weight to training sample:

$$W_{1,i} = \begin{cases} 0.5/m, i \leq m \\ 0.5/n, \text{otherwise} \end{cases} \tag{2.1}$$

   where m and n are the positive and negative pictures individually and $l = m + n$
3. For $t = 1$ to T
   Standardization of weight

$$W_{t,i} = W_{t,i} / \sum_{j=1}^{L} W_{t,j} \tag{2.2}$$

4. For each element j a frail classifier h_j is prepared and its mistake $\in_j$ is accessed w.r.t. $W_{t.}$

$$\varepsilon_j = \sum_{i=l}^{L} W_{t,i} * |h_j(X_i) - Y_i| \tag{2.3}$$

$$h_j = \begin{cases} 1 & p_j g_j(X) < p_j \theta_j \\ 0 & Otherwise \end{cases}, \tag{2.4}$$

   where $p_j \in (1, -1)$ demonstrates an equality bit and θ_j shows limit.
5. h_j is the classifier with least mistake rate of ε_t.
6. Updating the weight:

$$W_{t+l,i} = W_{t,i} * \beta_t^{1-e_i} \tag{2.5}$$

   If X_i is classified properly then $e_i = 0$ otherwise $e_i = 1$ and $\beta_t = (\varepsilon_t) / (1 - \varepsilon_t)$.
7. Final classifier

$$H_{(x)} \begin{cases} 1 \; if \sum_{t=1}^{T} a_t h_t(X) \geq 0.5 \sum_{t=1}^{T} a_t, \\ 0 \qquad \text{otherwise} \end{cases} \tag{2.6}$$

where $a_t = \log(\frac{1}{\beta_t})$.

2.2.1.1.3 Cascaded Detector

Owing the greedy characteristics, the enormous image can also be handled by the AdaBoost algorithm. In facial recognition, huge training set of images are explored to increase the computational efficiency and sequence of increased in the complicated classifiers so it is called as cascade and then gradually decreases in the false result (Fig. 2.4).[13,14]

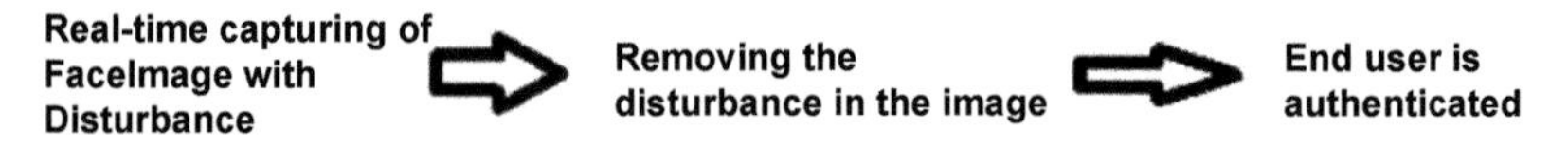

FIGURE 2.4 Cascade process.
Source: Ref. [14].

2.2.1.2 ADVANTAGES OF FACIAL RECOGNITION

- Facial recognition provides a high level of security against fraud detection.
- Accuracy rate is normally high in facial recognition.
- Facial recognition is one of the fully automated biometric systems.
- An individual can be easily identified among the group of people.
- Less time is used for the identification and recognition in facial recognition.

2.2.1.3 DISADVANTAGES OF FACIAL RECOGNITION

- Data storage and data processing are the two difficulties in facial recognition.
- Facial recognition is based on the image size, quality, camera angle, and face expression.

2.2.1.4 CHALLENGES OF FACIAL RECOGNITION

- Variation of face poses.
- Changes of facial expression over time.
- Variation in illumination condition.
- Resolution of image, image quality, and clear image.

2.2.2 FINGERPRINT RECOGNITION

Fingerprint recognition is one of the most commonly used biometric system methods for the identification and recognition of every individual. Due to the development in recent technology, fingerprint recognition system became small and low-cost when compared with other biometric systems. As a result, fingerprint-recognition system is implemented in the real-time applications, like mobile phones, laptops, buildings, and high-security purpose in military application. Fingerprint biometric system comprises scanner and processor, which stores, read, compare, and matches the scanned image against the stored image in the database. Ridges and valleys are two detailed pixel information of the fingerprint. Touch sensor and swipe sensor are the two categories under the fingerprint sensor based on the shape and size. In touch sensor, user places and holds the finger on the surface of the sensors and captured images transferred from the biometric system compares itself with the stored images in the databases and then it recognizes or authenticates an individual. In a swipe sensor, the user swipes vertically on the surface of the sensor, which compares and then authenticates the individual. The most commonly used sensor in fingerprint biometric system is a touch sensor because it authenticates user very fast compared with other sensors. There are mainly three types of classes in fingerprint recognition, they are arches, loops, and whorls.[3,5]

2.2.2.1 ARCHES

Arches are one of the simplest types of fingerprints that are composed of ridges that start from one end and it ends at another end as shown in Figure 2.5. Again, the arch is classified into two categories, that is, plain arch and tented arch. In plain arch ridges enter from one side and exit in other end but tented arches have spike in the center as shown in the figure.[9,10]

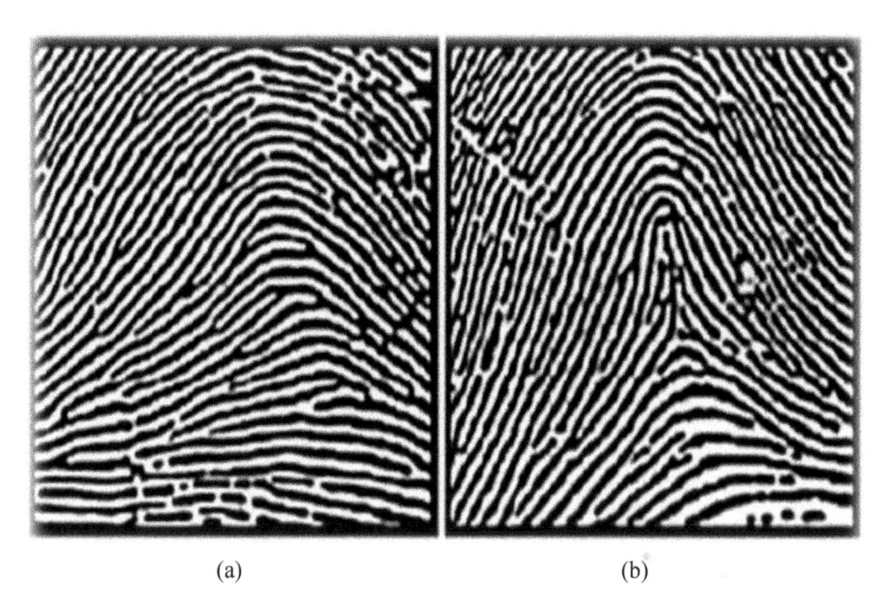

(a) (b)

FIGURE 2.5 (a) Plain arch and (b) tented arch of the fingerprint.

2.2.2.2 LOOP

The loop should have one delta and one or more ridges that start and ends with one side. These patterns are named based on the position that can be identified with the span and ulnar bones as an outspread circle or ulnar circle as shown in Figure 2.6.

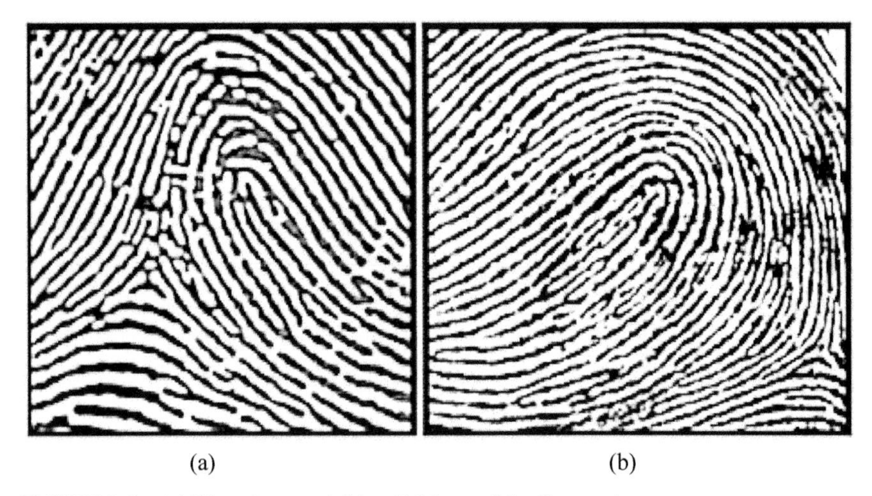

FIGURE 2.6 (a) Ulnar loop and (b) radial loop of the fingerprint.

2.2.2.3 WHORLS

Whorls have no less than one edge that makes to one finish circuit and consist of two or more deltas as shown in Figure 2.7. There are four types of whorls: central pocket whorl, plain whorl, accidental whorl, or double whorl.

Detailed flowchart for fingerprint identification, recognition, and authentication processes is shown in Figure 2.8.

2.2.2.4 ADVANTAGES OF FINGERPRINT RECOGNITION

- It is one of the unique identities of every individual.
- It uses less storage space.
- It has a very high accuracy rate.
- It is simple and secure to use for all including illiterate person also.

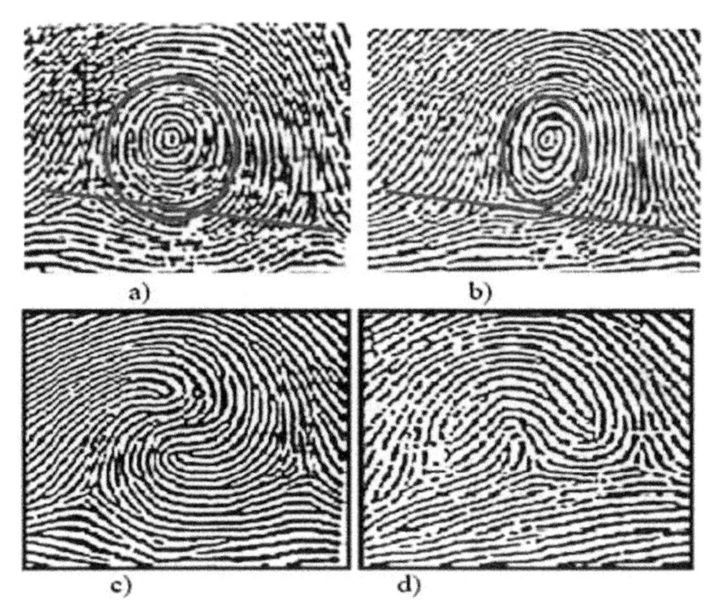

FIGURE 2.7 (a) Plain whorl, (b) central pocket whorl, (c) double whorl, and (d) accidental whorl, respectively.

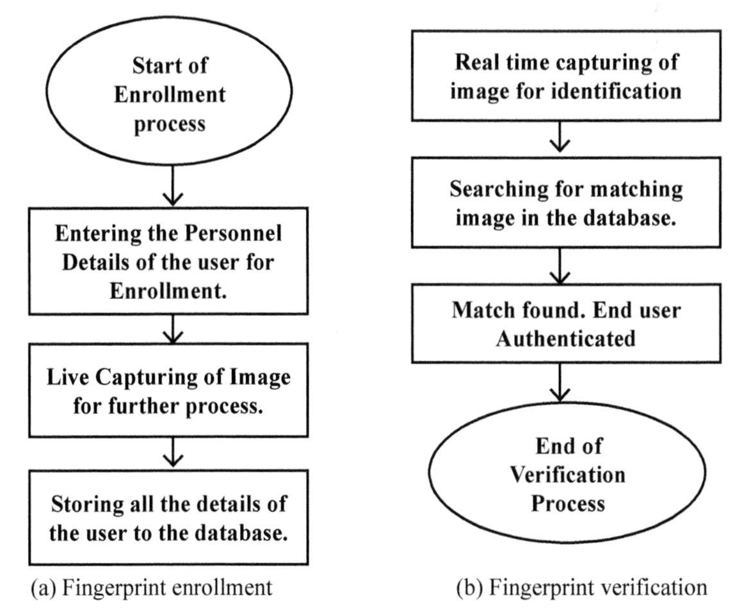

FIGURE 2.8 Fingerprint identification, recognition, and authentication processes.

2.2.2.5 DISADVANTAGES OF FINGERPRINT RECOGNITION

- It is unable to identify a person with and dryness, wetness or dirty of the finger's skin.
- Using cheap quality of components there is the possibility of unauthorized person access.

2.2.2.6 CHALLENGES OF FINGERPRINT RECOGNITION

- There is a possibility of mismatching due to physical distortion like finger injuries.
- Incorrect placement or movement of the finger over the sensor during the scanning process.
- Using finger plasticity or clay printing there is a chance of unauthorized access.
- Security and protection of the personal information.

2.2.3 IRIS RECOGNITION

Iris recognition (Fig. 2.9) is one of the techniques of a biometric system that works with mathematical pattern recognition on images or video images of every individual. Complex pattern of iris recognition is unique, secure, stable, and fast access when it is compared with other modalities. Iris recognition system captures a high resolution image of an individual to maintain the quality of the image and also for the fast recognition purpose.[8] The captured images are compared with the predefined images stored in the database. Here pattern recognition technology used to read and match the captured image and the image that is stored in the databases. There are mainly three steps in iris recognition for the identification, recognition, and authentication of an individual: live image capture, identifying the iris and optimizing the image, and comparing or matching the captured image with the stored images in the database.[6,7]

2.2.3.1 ADVANTAGES OF IRIS RECOGNITION

- Identification and Recognition of an individual within a fraction of seconds or less than 5 s.
- The accuracy rate is very high.

- It is also identified in case of with or without glass or contactless of the eyes.
- The degree of randomness of iris pattern is high which makes every individual's iris unique.

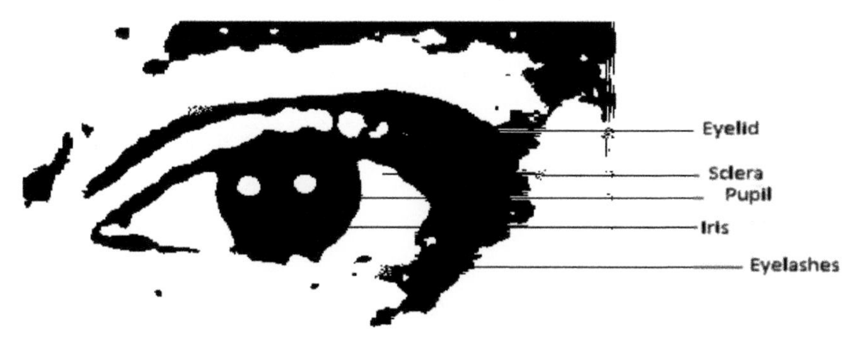

FIGURE 2.9 Represents the iris image.

2.2.3.2 DISADVANTAGES OF IRIS RECOGNITION

- It is very expensive.
- It requires a large amount of large storage space for the data storage purpose.
- It is unable or difficult to identify in the presence of refection.
- Distance between the eyes (iris) and the camera is limited for the identification and recognition.

2.2.3.3 CHALLENGES OF IRIS RECOGNITION

- Eye image should be cropped around the pupil.
- Crop eye image around the iris part.
- Finding the iris center and iris boundary.

2.2.4 PALM VEIN RECOGNITION

The palm vein acknowledgment is a standout amongst the most secure biometric system, used for the identification and recognition of an individual. Palm vein recognition is most secure and very difficult to duplicate because the veins are situated on the inner layer of the external skin. Palm vein recognition biometric system authenticates an individual by the physical characteristic that is the organization of the palm vein. Geometric, cosine similarity, and wavelet are the various features of the palm vein that are used for the identification and recognition of an individual (Fig. 2.10). Compared with other biometric methods, palm vein recognition is, generally, one of the secure, complex, and very difficult to duplicate or forge. Near-infrared illumination technology is used to capture the palm vein of an individual and because of less sensitivity to the temperature and humidity the image will be captured and displayed clearly (Fig. 2.11).[1]

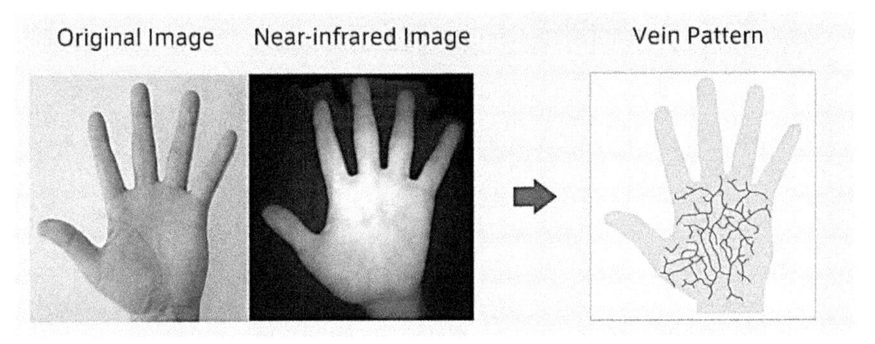

FIGURE 2.10 Palm vein image.

Source: Ref. [18].

2.2.4.1 *APPLYING HYBRID PRINCIPAL COMPONENT FOR PALM VEIN RECOGNITION*

In palm vein acknowledgment cross breed central segment investigation and self-sorting out map (sop) are utilized to distinguish the accuracy of the captured image of the palm. The Principal Component Analysis and Artificial Neural Network (PCA-ANN) considered two times when inputs to ANN that is for unscaled and scaled. Unscaled are the raw scale between 0 and 255 and scaled are scale between 0 and 0.9. The operation of the system can be carried out based on different image resolution, training sets of images, acknowledgment time, and precision. The unscaled and scaled PCA-ANN gives optimal recognition accuracy that is in between 55% and 98% for unscaled and 56% and 99% for scaled, respectively, at resolution of between 30×30 and 60×60 pixels level of cropping. The error rates are also calculated to know the efficiency of the algorithm. The false acceptance rate (FAR) is in between 2.5% to12.5% for unscaled and 2.5% to 15% for scaled. False rejection rate (FRR) will be between 2% and 82% for unscaled and 1% and 81% for scaled at 0.0001 limit separately. Equal error rate (EER) was 9% for unscaled PCA-ANN at 49 pixels and 12% for the scaled PCA-ANN nearly 46-pixel determination. So that lower pixel resolution can be obtained by scaled (46.37) PCA-ANN and by large framework precision will be achieved by the scaled PCA-ANN (46.37) than unscaled PCA-ANN (49.53).[18]

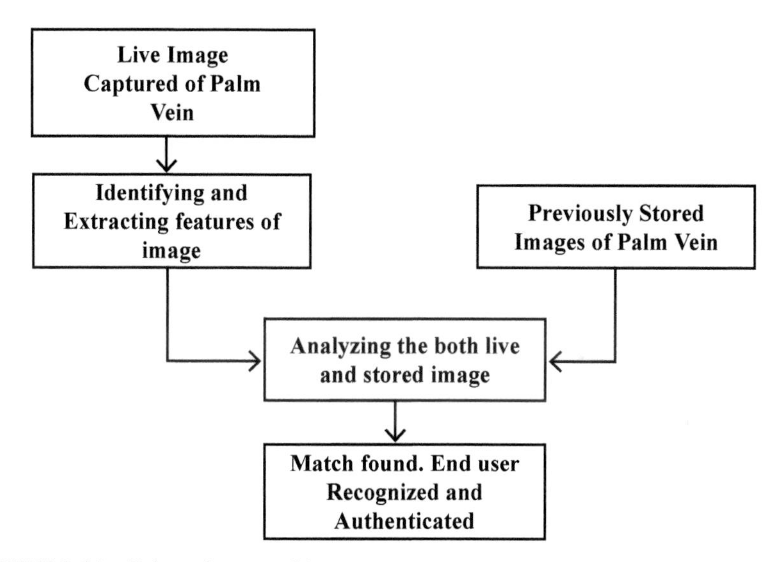

FIGURE 2.11 Palm vein recognition process.

Source: Ref [18].

2.2.4.2 ADVANTAGES OF PALM VEIN RECOGNITION

- Compared with other biometric systems, it gives more accuracy.
- Reading, comparing, and matching performance is very high.
- Easy to use and maintain.
- Security level and accuracy level is very high.

2.2.4.3 APPLICATIONS OF PALM VEIN RECOGNITION

- For management in healthcare organization.

- For the identification and recognition of the doctor to access the patient health information or patient records that are stored under high security.
- For the identification and authentication of the patient in the hospitals.
- Attendance authentication system.
- It is used as an attendance system in various government offices, schools, colleges, institutions, and so on.
- It automatically calculates the exact in and out times of an individual.

2.2.4.4 DISADVANTAGES OF PALM VEIN RECOGNITION

- It requires a large amount of space to store the image.
- The cost is high compared with other biometric systems.
- Security and accuracy level are very high.
- Very slow to authenticate an individual compared with other biometric systems.

2.3 CONCLUSIONS

From this, we finally concluded that important information can be protected using the different biometric systems and can be accessed by the authenticated user. Here, recent technologies, features, application, advantages, and disadvantages are explained so that it can be easy to know which biometric system that is suitable for our application.

KEYWORDS

- **fingerprint recognition**
- **facial recognition**
- **iris recognition**
- **palm vein recognition**
- **AdaBoost algorithm**
- **principal component analysis**

REFERENCES

1. Thakkar, D. Biometric Technology, Facial Recognition, Finger Vein Recognition, Fingerprint Recognition, Iris Recognition, Palm Vein Recognition, Voice Recognition. **2018**. https://www.bayometric.com/biometrics-face-finger-iris-palm-voice/.
2. Tenuta, A. Benefits of Facial Recognition Time Tracking Solutions. *EPAY Systems* **2017**. http://blog.epaysystems.com/5-benefits-of-facial-recognition-time-tracking-solutions.
3. Agarwal, T. Fingerprint Identification. *ElProCus* **2017** https://www.elprocus.com/fingerprint-identification/.
4. Advantages and Disadvantages of Recent Technologies of the Biometric System. *Biometrics PbWorks*, **2011**. http://biometrics.pbworks.com/w/page/14811349.
5. Soffar, H. Fingerprint Scanners Uses Features, Advantages and Disadvantages. 2018. https://www.online-sciences.com/technology/fingerprint-scanners-uses-features-advantages-and-disadvantages/.
6. Daugman, J. *Advantages and Disadvantages of Iris Recognition,* Biometric Today and University of Cambridge. 2018. https://biometrictoday.com/25-advantages-dis-advantages-iris-recognition/.
7. Ranjan, S.; Sevugan, P.; Swarnalatha, P.; Gopu, M.; Sundararajan, R.; IRIS Recognition System. Int. Res. J. Engineer. Technol. **2017,** *4* (12), 864–868.
8. Kumar, M.; Priyanka Fingerprint Recognition System: Issues and Challenges. *Int. J. Res. Appl. Sci. Eng. Technol.* **2018,** *6* (II), 556–561.
9. Krishnasamy, P.; Belongie, S.; Kriegman, D. *Wet Fingerprint Recognition: Challenges and Opportunities*, International Joint Conference on Biometrics (IJCB), University of California, San Diego, 2011.
10. Dhivakar, B.; Sridevi, C.; Selvakumar, S.; Guhan, P. *Face Detection and Recognition Using Skin Color*, 3rd International Conference on Signal Processing, Communication, and Networking (ICSCN), Chennai, India, 2015.
11. Tu, Y.; Vi, F.; Chen, G.; Jiang, S.; Huang, Z. *Fast rotation invariant face detection in color image using multiclassifier combination method,* International Conference on E-health Networking, Digital Ecosystem and Technologies, Shenzhen, China, 2010, pp 212–218.
12. Tofighi, A.; Monadjemi, S. A. *Face Detection and Recognition Using Skin Color and Adaboost Algorithm Combined with Gabor Feature and SVM Classifier,* International Conference on Multimedia and Signal Processing, Guilin, Guangxi, China, 2011, pp 141–145.
13. Mohanty, R.; Raghunadh, M. V. *A New Approach to Face Detection based on YCgCr Color Model and Improved AdaBoost Algorithm*, International Conference on Communication and Signal Processing, Melmaruvathur, India, 2016.
14. Han, X.; Du, Q. *Research on Face Recognition Based on Deep Learning*, Sixth International Conference on Digital Information, Networking, and Wireless Communications (DINWC), Beirut, Lebanon, 2018.
15. Raghavendra, C.; Kumaravel, A.; Sivasubramanian, S. *Iris Technology: A Review On Iris Based Biometric Systems For Unique Human Identification,* International

Conference on Algorithms, Methodology, Models and Applications in Emerging Technologies (ICAMMAET), Chennai, India, 2017.

16. Thakre, S.; Gupta, A. K.; Sharma, S. *Secure Reliable Multimodal Biometric Fingerprint and Face Recognition*, International Conference on Computer Communication and Informatics (ICCCI), Coimbatore, India, 2017.
17. Sasidharan, S.; Azath, M. Study on Palm Vein Authentication. *Int. J. Sci. Res.* **2015,** *3* (1), 28–32.
18. Chowdhary, C. L. Appearance-based 3-D Object Recognition and Pose Estimation: Using PCA, ICA and SVD-PCA Techniques. *LAP Lambert Acad.* **2011.**
19. Chowdhary, C. L.; Acharjya, D. P. Singular Value Decomposition–Principal Component Analysis-Based Object Recognition Approach. In *Bio-Inspired Computing for Image and Video Processing;* 2018; p 323.
20. Chowdhary, C. L. Application of Object Recognition with Shape-Index Identification and 2D Scale Invariant Feature Transform for Key-Point Detection. *Feature Dimension Reduction for Content-Based Image Identification;* 2018; pp 218–231.
21. Chowdhary, C. L.; Muatjitjeja, K.; Jat, D. S. *Three-dimensional Object Recognition Based Intelligence System for Identification.* International Conference on Emerging Trends in Networks and Computer Communications (ETNCC), 2015.
22. Chowdhary, C. L.; Ranjan, A.; Jat, D. S. Categorical Database Information-theoretic Approach of Outlier Detection Model, *Annals*. Computer Science Series. 14th Tome 2nd Fasc. **2016,** 29–36.
23. Chowdhary, C. L. Linear Feature Extraction Techniques for Object Recognition: Study of PCA and ICA. *J. Serbian So. Comput.Mech.* **2011,** *5* (1), 19–26.

CHAPTER 3

Speech-Based Biometric Using Odia Phonetics

BASANTA KUMAR SWAIN[1*] and SANGHAMITRA MOHANTY[2]

[1]*Department of Computer Science and Engineering, Government College of Engineering, Kalahandi, Bhawanipatna, India*

[2]*Department of Computer Science and Application, Utkal University, Bhubaneswar, India*

Corresponding author. E-mail: technobks@yahoo.com

ABSTRACT

In this chapter, the focus is on behavioral approach, that is, speech signal processing for the user authentication, which seems to be a more complicated task in comparison to any physiological biometric technique. The research work is carried out on text-dependent voice biometric system by hybridizing with a speech recognition engine for enhancing the performance of the biometric identification system. The key reason behind the conglomeration of automatic speech recognition engine with the speaker recognition system is that it will fetch the information in physiological domain relating to the shape of the vocal tract (pharynx, larynx, mouth, jaw, tongue, and lungs) of an individual as well as the behavioral aspects, that is, place and manner of articulation.

3.1 INTRODUCTION

Speech-based biometric for user authorization is a task of performing speaker's authentication in a hassle-free, convenient, secure, and robust way. The use of biometric systems is rapidly increasing day by day in every corner of society due to the advancement of information technology services.[1,2] In the current market scenario two types of biometric systems are available, namely, physiological and behavioral. The physiological biometric system is based on fingerprint, iris, hand, face, etc. On the other hand, behavioral system is based on voice, signature, gait, etc. The biometric-based authentication of user is highly reliable than the traditional "code or password" based authentication techniques as the later authentication process have innumerable problems, that is, memorizing, fear of losing or hacking, etc.[3,4] Hence, the biometric-based authentication is a lucrative and better option for the recent day.

The first attempt for the development of automatic speaker recognition was made by Pruzansky at Bell Labs during early 1970s based on spectrogram similarity measurement. Furui at Bell Labs proposed an experimental system by using cepstral coefficients and their first and second polynomial coefficients to work in dialed-up telephone lines. Researchers of International Business Machines Corporation also developed a speaker recognition system by applying linear discrimination analysis in the intervening period of time. In 1980s, the main focus was given over statistical approaches, like vector quantization (VQ) or Hidden Markov Model (HMM) for the development of the speaker recognition system. But in the 1990s, the center of attention was on the robustness of the speaker identification system. Matsui et al[5] made analysis over VQ and the discrete/continuous ergodic HMM-based method from the point of view of robustness against variation of utterance duration. However, in the 21st century, the research over speaker recognition is going on by adopting the multitudes of strategies, such as imposter handling, noise robustness, minimization of intraspeaker variation, etc., based on high-level features, score normalization techniques, etc.

In this chapter, the focus is on the behavioral approach, that is, speech signal processing for user authentication that seems to be a more complicated task in comparison to any physiological biometric technique. The research work is carried out on text-dependent voice biometric system by hybridizing with a speech recognition engine for enhancing the performance of the biometric identification system. The key reason behind the conglomeration of automatic speech recognition engine with speech recognition system is that it will fetch the information in physiological domain related to the shape of vocal tract (pharynx, larynx, mouth, jaw, tongue, and lungs) of an individual as well as the behavioral aspects, that is, place and manner of articulation.

This chapter is organized as follows: Section 3.2 illustrates about voice biometric speech corpus, Section 3.3 emphasizes on feature extraction, Section 3.4 depicts the development of speech based biometric in conglomerate approach, Section 3.5 demonstrates the experimental setup and results, and the last section says about the conclusions and future work.

3.2 VOICE BIOMETRIC SPEECH CORPUS

In this chapter, researchers have demonstrated the result of voice-based biometric system by using self-created Odia spoken corpus. The contents of speech corpus consist of isolated words or digits, continuous digits of length 6, prompted texts, unique phrase, and client-dependent pass-phrase, etc. Speech corpus is representing around 107 numbers of male and female speakers, having age between 18 and 55 years old. In recording sessions, volunteer speakers are seated at a table with noise cancellation; unidirectional headphone connected to laptops or desktops and required to speak into the headphone. The student's computer laboratories, staff personal chambers, and residence as well are used as recording environment for spoken data collection. The efforts are made to engage different manufactures computers and recording devices during speech corpus collection. Speech files are collected from the same speaker in multiple times, but in different sessions, in order to handle the speech variability.[7] The rerecordings of voice passwords are also considered on the demand of session chair personnel or speaker's request. The Praat software is used to capture and manage the speech received over the microphone.[8] Speech files are recorded at 16,000 Hz, 16-bit, and mono channel. The recorded file is stored in wave format. Table 3.1 describes the biometric speech corpus.

TABLE 3.1 Speech Corpus for Speaker Verification.

Gender	No. of speakers	Nature of voice password	Property of Voice password
Male	82	Isolated words/continuous digits/prompted texts/unique phrase/pass-phrase	Mono, 16 bit, 16,000 Hz, and PCM
Female	25	Isolated words/continuous digits/prompted texts/unique phrase/pass-phrase	Mono, 16 bit, 16,000 Hz, and PCM

3.3 FEATURE EXTRACTION

The developed speech corpus is then exposed to the feature extraction component. Feature extraction is meant for parameterization of spoken files. The users need to try considering the properties namely universality, distinctiveness, permanence, collectability, performance, acceptability, circumvention, etc., during feature extraction as the accuracy of performance of the biometric system will depend greatly on training and testing feature vectors. In this research, there is the utilization of two feature parameters, namely, Mel-frequency cepstral coefficients (MFCC) and their variants as well as gammatone frequency cepstral coefficients (GFCC).

MFCCs are calculated by applying a series of mathematical transformations over voice signal. Fast Fourier transform (FFT) is applied over each 10 ms frame with overlapping analysis window of around 25 ms that generates the power spectra and then it is passed series of rectangular "pass band" filters. In each frequency band, the energy is log-compressed and discrete cosine transform (DCT) is applied to obtain the cepstral parameters.[9,10]

A Mel is a unit of the pitch where frequencies obtained by FFT below around 1000 Hz are mapped linearly to the Mel scale, and those above 1000 Hz are mapped logarithmically according to the eq 3.1.

$$mel(f)=1127\, ln\left(1+\frac{f}{700}\right). \tag{3.1}$$

The resulting coefficients are uncorrelated due to the application of logarithmic function and, moreover, those uncorrelated coefficients are leading to generate better acoustic model by machine learning algorithms.

In addition to the spectral coefficients, first order (delta) and second-order (delta–delta) regression coefficients are appended.

If the original (static) feature vector is y_t, then the delta parameter, Δy_t^s, is given by

$$y_t^s = \frac{\sum_{i=1}^{n} w_i \left(y_{t+1}^s - y_{t-i}^s\right)}{2\sum_{i=1}^{n} w_i^2}, \tag{3.2}$$

where n is the window width and w_i are the regression coefficients.

The delta–delta parameters, $\Delta^2 y_t$, are derived in the same fashion, but using differences of the delta parameters.[11,12] When concatenated together these form the feature vector y_t,

$$y_t = [y_t^{s\,T} \quad \Delta y_t^{s\,T} \quad \Delta^2 y_t^{s\,T}]. \tag{3.3}$$

Another feature parameters are extracted from speech files called GFCC. The GFCC values are calculated by passing through the input signal to a 64 channel gammatone filter bank and then the filter response is rectified and decimated to 100 Hz as way of time windowing. It results a time-frequency representation that is a variant chochleargram. The resultant time-frequency is exposed to cubic root function. Finally, DCT is applied to derive the cepstral features.[13]

The filter is defined in the time domain by the following impulse response:

$$h(t) = t^{a-1} \quad e^{-2\pi bt} \cos(2\pi f_c t + \varphi) \tag{3.4}$$

where a, b indicates the filter, t is time, f_c is the filter's center frequency, and φ is the phase.

GFCC is based on equivalent rectangular bandwidth (ERB) frequency scale that produces the finer resolution at low frequencies.[14]

Glasberg and Moore have represented human data on ERB of the auditory filter with a function as follows:

$$\text{ERB}\,(f_c) = f_c \,/\, \text{Q} + B_0 \tag{3.5}$$

$$B_0 = 24.7,\ \text{Q} = 9.64498,$$

where B_0 is minimum bandwidth and Q is the asymptotic filter quality at large frequencies.

3.4 DEVELOPMENT OF SPEECH-BASED BIOMETRIC IN CONGLOMERATE APPROACH

This chapter is focused on the development of speech-based biometric by conglomerating two distinct speech processing techniques namely speech recognition and speaker identification. The performance of speech-based biometric system is enhanced due to the hybridization of speech recognition engine. This work has used speech recognizer engine as a preprocessing module for speaker identification. In the recent market, most of the researchers are using pattern-recognition techniques for the development of speaker identification system. But the loophole of any pattern recognition technique is that it demands huge time for training model generation from voluminous input data. So, this approach will act as a filtering technique that will generate the training model in a very short span of time by using only the relevant spoken data and discarding the unnecessary data for that moment. This research work has developed the speech recognizer engine that is based on Odia language phonetics. But this approach can be extended to develop a speech recognizer engine of any language.

This speech recognition engine is based on HMM machine learning technique.

3.4.1 HMM AS SPEECH RECOGNIZER ENGINE

The voice passwords are represented in terms of fixed size acoustic vectors $Y = y_1, \cdots, y_T$. The decoder then attempts to find the sequence of words $W = w_1, \cdots, w_K$ which is most likely generated from Y.

$$\widehat{W} = argmax\,[p(W|Y)]. \tag{3.6}$$

Bayes' rule is used to transform the above eq 3.6 into the equivalent problem of finding:

$$\widehat{W} = \underset{W}{argmax}[p(Y|W)p(W)]. \tag{3.7}$$

The likelihood $p(Y|W)$ is determined by an acoustic model and the prior $p(W)$ is determined by a language model.[15,16]

The spoken words, W, are represented in sequence-based phones. The likelihood $p(Y|W)$ can be computed over multiple pronunciations for possible pronunciation variation of W as follows,

$$p(Y|W) = \sum_{Q} p(Y|Q)\,p(Q|W), \tag{3.8}$$

where Q is a particular sequence of pronunciations and q represents individual base phone.[17,18]

Figure 3.1 illustrates how states and transitions are carried out in an HMM to represent phonemes, words, and sentences. The states are interpreted as acoustic models in HMM-based speech recognition indicates the sounds that will be heard during their corresponding segments of speech.[19,20] As human voice always goes forward in time, transitions in speech applications always go forward or make a self-loop.

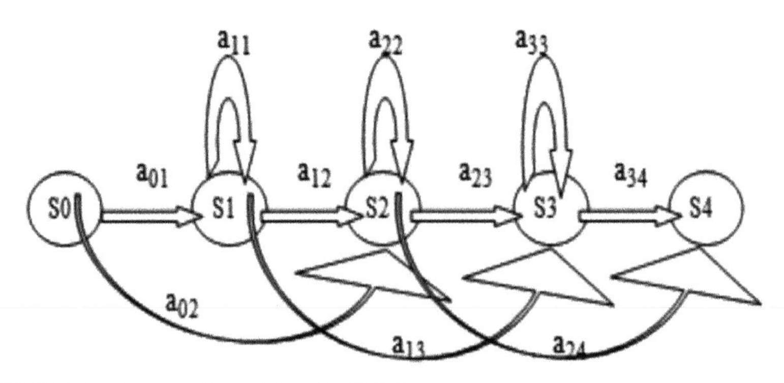

FIGURE 3.1 States and transitions in HMM.

The probability of making a particular transition from state si to state sj is given by the transition probability a_{ij}.

Speech recognizer accuracy is greatly improved by taking the advantage of language model that gives the priori information *p*(*W*) on the possible sequences to be recognized.[21–40] The language model is treated as human comprehension mechanism of spoken utterances. A person recognizes the meaning of a mispronounced word from the part of speech. There are different types of language models used in the speech recognition engine namely uniform models, Finite state language models, stochastic models, etc.[24–29] But this research article is used stochastic-based language model that calculates the joint probability of a word and its preceding words. *p*(*W*) can be decomposed into the product of the word probabilities as

$$p(W)=P(w_1)P(w_2|w_1)P(w_3|w_1w_2).........P(w_n \mid w_1w_2......w_{n-1})$$
$$\prod_{i=1}^{n} P(w_i \mid w_1w_2......w_{i-1}). \tag{3.9}$$

The above equation- tractable is using trigram language model.

$$p(W)=P(w_1)P(w_2|w_1)P(w_3|w_2w_1).........P(w_n \mid w_{n-1}w_{n-2})$$
$$\prod_{i=1}^{n} P(w_i \mid w_{i-1}w_{i-2}). \tag{3.10}$$

Probability is estimated using the relative frequency approach[28,29]:

$$P(w_i = w^m \mid w_{i-1} = w^{m'}, w_{i-2} = w^{m''})$$
$$= N\left(\frac{w_i = w^m \mid w_{i-1} = w^{m'}, w_{i-2} = w^{m''}}{w_{i-1} = w^{m'}, w_{i-2} = w^{m''}}\right) \tag{3.11}$$

Once the speech recognition engine is ready, it can be linked to the speaker identification module. The hybridization of speech recognizer with speaker identification techniques gives the immense benefit of pruning. There is a risk of over-fitting of the training data as well as poorly generalizing to the new samples for big size database. Pruning will reduce the size of training files by selecting the required samples from the original training database. Ultimately, the recognition engine reduces complexity of final classifier and improves the predictive accuracy at the end. The holistic algorithm is represented behind the speech based biometric system as follows:

3.4.2 ALGORITHM OF CONGLOMERATION

Step 1: Generate the MFCC of Unknown speaker's voice file which will be authenticated.

Step 2: Decode the Unknown speaker's voice file using Oriya automatic speech recognition (ASR).

Step 3: Find the percentage of matching of decoded result with voice password of different speakers that are collected during the enrolment phase.

Step 4: Select the voice password files from biometric speech corpus those are having a percentage of matching more than a specific threshold.

Step 5: Extract the GFCC feature parameters from the selected voice password files.

Step 6: Now use the support vector machine (SVM)-based speaker identification technique to identify the unknown speaker.

There is a written a program that will find edit distance between the speech recognizer result (predicted password) and transcription file (contains the real voice password).The program works as follows:

- Intialize INS_COST=1, DEL_COST=1 and REP_COST=1
- int editDist(char str1[], int s1, char str2[],int s2)

```
{
For i = 0 to s1-1
        {
        For j=0 to s2—1
        {
                If (i==0)
                        Cost[i][j]=j*INS_COST
                Else if(j==0)
                        Cost[i][j]=i*DEL_COST
                      Else
                        Cost[i][j] = -1
                }
        }

        For i = 0 to s1-1
                {
                For j=0 to s2—1
                        {
                        x= Cost[i-1][j] + DEL_COST
```

```
                y= Cost[i][j-1] + INS_COST
                bi=((str1[i-1] != str2[j-1])? 1 : 0)
                z=cost[i-1][j-1] + bi* REP_COST
                cost[i][j] = min(x, min(y, z))
        }
    }
    Return cost[size -1][ size2 - 1]
```

The output of Step 3 represents the core values in terms of percentage matching between predicted passwords and real passwords. Step 4 uses the output of Step 3 as input parameter. The main principle of this step is selection of voice password files from whole spoken corpora for Step 5 that will be used as the training database for SVM classifier. Selection of voice files from voluminous biometric speech corpus is based on prespecified threshold. Step 4 acts pruning because it selects the voice files for next step that seem suitable on the basis of score value more or equal to the threshold and discards the voice files that are irrelevant with respect to test voice file (score value less to threshold).

Step 5 relates to the extraction of GFCC values from selected voice files that are elaborately discussed in Section 3.3.

Step 6 deals with SVM classifier for speaker identification. The SVM classifier is trained using the feature parameters extracted in Step 5. The detailed working procedure of SVM classifier is illustrated in the following section.

3.4.3 SVM CLASSIFIER BASED ON PRUNED SPEECH CORPORA

The SVM classifier is based on risk minimization (RM) mechanism by adopting the concept of duality theory of optimization to generate computationally tractable model parameters in higher-dimension feature space. The key idea behind the RM is reducing the misclassification error with the help of the maximal margin technique. This algorithm works very efficiently for high-dimensional feature vectors size.[30,31]

SVM constructs a set of hyperplanes in higher-dimensional feature space that are at largest distance from training samples of different classes. Hyperplane is represented as a line for two-dimensional cases and a plane for three-dimensional scenarios.[32,34]

A hyperplane in two-dimension is defined in terms of parameters β_0, β_1, and β_2 as an equation of line

$$\beta_0 + \beta_1 X_1 + \beta_2 X_2 = 0. \tag{3.12}$$

Equation 3.12 can be extended to n-dimensional from two dimensional as follows:

$$\beta_0 + \beta_1 X_1 + \beta_2 X_2 + \ldots\ldots\ldots + \beta_n X_n = 0 \tag{3.13}$$

An object, X may also lie in either side of the hyperplane that can be represented as

$$\beta_0 + \beta_1 X_1 + \beta_2 X_2 + \ldots\ldots\ldots + \beta_n X_n < 0 \tag{3.14}$$

$$\beta_0 + \beta_1 X_1 + \beta_2 X_2 + \ldots\ldots\ldots + \beta_n X_n < 0 \tag{3.15}$$

A test observation can be classified on the basis of which side it lies. Figure 3.2 represents two classes of the biometric system with a maximal marginal hyperplane.

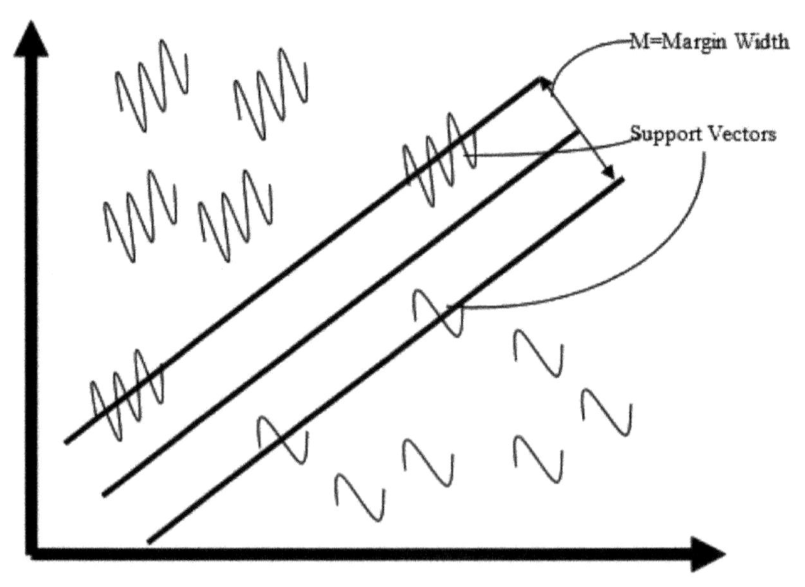

FIGURE 3.2 Two classes of biometric system with large margin.

The optimal hyperplane always tries to correctly segregate most of training instances but it may incorrectly classify a few instances of a given set. SVM is treated as an optimization problem

$$\text{maximize} \quad M \tag{3.16}$$

$$\beta_0, \beta_1, \beta_2, \ldots\ldots\ldots \beta_n, \in_1, \ldots\ldots \in_n, \tag{3.17}$$

$$\text{Subject to } \sum_{j=1}^{n} \beta_j^2 = 1, \tag{3.18}$$

$$+y_i\,(\beta_0 \beta_1 X_1 + \beta_2 X_2 + \ \ldots\ldots + \beta_n X_n) \geq M(1 - \in_i) \tag{3.19}$$

$$\in_i \geq 0, \sum_{i=1}^{n} \in_i \leq C, \tag{3.20}$$

where C is a nonnegative tuning parameter. M represents the largest margin and $\in_1$, $\in_2$ ..., $\in_n$ are slack variables that allow few observations to be on the improper side of the hyperplane.[35,36]

The SVM classifier is also capable of creating nonlinear decision boundaries to classify nonlinearly separable data in feature space using Kernel functions magic. In this chapter, we have used Radial Basis Function (RBF) as kernel function to separate voice passwords vectors in nonlinearly.[37–45] RBF function is defined as

$$K\left(x, x_{i,}\right) = \exp(-\gamma \,||X - X_i)\,||^2\ ). \tag{3.21}$$

3.5 EXPERIMENTAL SETUP AND RESULTS

In this chapter, SPHINX 3-based and HMM-based speech recognizer are the most suitable continuous and isolated voice passwords in Odia language. But approach followed in this research chapter can be easily extended to any spoken language. The speech recognizer is trained using feature parameters MFCC and their variants of length 39. The vital components adopted in the development of speech recognition engine are discussed below.

3.5.1 TRANSCRIPTION FILE

After collecting the voice passwords from different users, it is used to prepare the transcription file. The transcription file represents the text version of audio files present in the biometric speech corpus. It is prepared manually by looking into the content of individual speech file. It also represents noise as well as silence zone of voice files. Sample entries of the transcription file are shown in Figure 3.3.

```
<s> tini chari na atha  </s> (satnami3498_pwd2)
<s> tini chari na atha  </s> (satnami3498_pwd3)
<s> tini chari na atha  </s> (satnami3498_pwd4)
<s> tini chari na atha  </s> (satnami3498_pwd5)
<s> somya </s> (somya1)
<s> somya </s> (somya2)
<s> somya </s> (somya3)
<s> somya </s> (somya4)
<s> sata sata na na   </s> (subha7799_pwd1)
<s> sata sata na na   </s> (subha7799_pwd2)
<s> sata sata na na   </s> (subha7799_pwd3)
<s> sata sata na na   </s> (subha7799_pwd4)
<s> sata sata na na   </s> (subha7799_pwd5)
<s> surya eka pancha panchaasi  </s> (suryaekapancapanchaasi_pwd1)
<s> surya eka pancha panchaasi  </s> (suryaekapancapanchaasi_pwd2)
<s> surya eka pancha panchaasi  </s> (suryaekapancapanchaasi_pwd3)
```

FIGURE 3.3 Sample entries of transcription file.

3.5.2 DICTIONARY

The voice-based biometric is using Odia speech recognition engine in the preprocessing stage. There is a written program in Java that generates phone-based dictionary where each word is mapped to their string of phones by looking into the pronunciation of word. The English representations are used of Odia language phones. Example of self-created dictionary file is shown in Figure 3.4.

3.5.3 PHONEME FILES

Phoneme file contains exactly the same number of distinct phones with respect to the dictionary. It contains one phone in each line, and duplication of phone is prohibited. A java program is written that will preserve the distinct phones present in the dictionary and discard the redundant phones in phoneme files. Figure 3.5 represents the phoneme file format used in speech recognition engine.

```
panch         p aa n c
alisa         aa ll i sh aa
sunu          sh u n u
babu          b aa b u
babun         b aa b u n
bapi          b aa p i
namaskar      n m a s k aa r
dalija        d a ll i j aa
gudu          g u dd u
jajati        j a j aa t i
puri          p u r i
mahanadi      m a h aa n a d ii
krishna       k r i sh n aa
harekrushna   h a r e k r u sh n a
mama          m aa m aa
pintu         p i n tt u
rajendra      r aa j e n d r a
```

FIGURE 3.4 Phone-based dictionary.

```
c
ch
j
tt
tth
dd
t
d
n
p
b
m
-
```

FIGURE 3.5 Sample entries of phoneme file.

3.5.4 LANGUAGE MODEL

Speech recognition accuracy is enhanced by using a language model. In this research work, a statistical language model toolkit called CMUCLM is used. The trigram language model is applied here. Sample of Language model used in this chapter is shown in Figure 3.6.

```
odia.ug.lm
################################################################################
================================================================================
=============== This file was produced by the CMU-Cambridge ===============
===============    Statistical Language Modeling Toolkit    ===============
================================================================================
This is a 3-gram language model, based on a vocabulary of 41 words,
  which begins "</s>", "<s>", "alisa"...
This is an OPEN-vocabulary model (type 1)
  (OOVs were mapped to UNK, which is treated as any other vocabulary word)
Good-Turing discounting was applied.
1-gram frequency of frequency : 1
2-gram frequency of frequency : 2 2 1 21 39 17 1
3-gram frequency of frequency : 3 3 1 30 56 25 2
1-gram discounting ratios : 0.50
2-gram discounting ratios :
3-gram discounting ratios :
This file is in the ARPA-standard format introduced by Doug Paul.

p(wd3|wd1,wd2)= if(trigram exists)           p_3(wd1,wd2,wd3)
                else if(bigram w1,w2 exists) bo_wt_2(w1,w2)*p(wd3|wd2)
                else                         p(wd3|w2)

p(wd2|wd1)= if(bigram exists) p_2(wd1,wd2)
            else              bo_wt_1(wd1)*p_1(wd2)

All probs and back-off weights (bo_wt) are given in log10 form.
```

FIGURE 3.6 Sample of trigram language model.

The Sphinx 3 decoder is used for conversion of speech file into textual format. The decoder first uses the feature parameters (MFCC) from the test voice password and then uses trained model of HMM-based on prepared phoneme file, transcription file, dictionary, and language model.

The developed speech recognition engine performance is measured in terms of sentence error rate and word error rate that are found as 22.9% and 14.9%, respectively. So, another way that the accuracy rate of the ASR engine is 77.1% and 85.1% at the sentence level and word level, respectively.

As per Section 1.4.2, it is measured the variation between the predicted passwords of ASR engine and actual passwords used by users. A program is written that will find edit distance between the speech recognizer results (predicted password) and transcription file (contains the real voice password). A snapshot of the comparison result is shown in Figure 3.7.

```
25.00
0.00
0.00
100.00
0.00
0.00
25.00
0.00
-25.00
75.00
0.00
0.00
7.81
```

FIGURE 3.7 Sample values of output of edit distance program.

Step 4 of the algorithm is accomplished by a small written program that compares the Step 3 result, that is, edit distance with a threshold value. If edit distance is greater than or equal to the predefined threshold then that speaker ID is preserved in a file for the future course of action. Step 4 actually does the job of pruning of speaker database. In this research chapter, there is a setting of threshold value as 25.00.

Now the resultant spoken corpus as per the selected speaker ID will be used in training model generation for speaker identification. In this research article, the statistical pattern recognition technique is used and SVM is called as a classifier for speaker identification which uses GFCCs as feature parameters. But feature vectors (GFCC) are extracted from the voice files IDs returned by Step 4. Hence, it is good to avoid the feature extraction from whole speech corpus (that contains all voice passwords). Figure 3.8 represents the feature parameters (GFCC) relating to speakers selected on the basis of threshold value. These are used for training purpose by SVM classifier.

The first digit of every paragraph in Figure 3.8 represents speaker identity and remaining parameters numbered from 1 to 22 that represent GFCC values. There is a java program called "Feat2SVM" that will automatically attach value to the speaker ID and assign digits to each feature vector.

```
2 1:3.4574865e-01 2:3.4029081e-01 3:3.9539361e-01 4:3.7867790e-01 5:3.8746121e-01 6:4.1233866e-01 7:4.5773243e-01 8:3.8479104e-01 9:4.7362920e-01 10:5.7702964e-01
11:7.9682639e-01 12:8.9075599e-01 13:8.5908708e-01 14:7.3821527e-01 15:8.2392344e-01 16:8.3623650e-01 17:9.8712491e-01 18:9.1470454e-01 19:8.8163682e-01
20:9.8364232e-01 21:8.5077180e-01 22:9.3848879e-01
2 1:2.6609281e-01 2:4.2599172e-01 3:3.4202114e-01 4:4.1987966e-01 5:2.2951901e-01 6:6.6083309e-01 7:6.6409673e-01 8:9.0944722e-01 9:9.3139585e-01 10:9.3629969e-01
11:9.0657057e-01 12:8.1300171e-01 13:8.3308731e-01 14:8.9226384e-01 15:8.4300254e-01 16:8.9955555e-01 17:8.1776261e-01 18:9.8159414e-01 19:9.2070132e-01
20:7.7034441e-01 21:9.7858172e-01 22:8.4759707e-01

21 1:1.0727747e-01 2:1.1705867e-01 3:1.6069361e-01 4:1.9437360e-01 5:3.2427794e-01 6:4.2363505e-01 7:5.7360881e-01 8:9.1387339e-01 9:9.0190357e-01 10:9.9567825e-01
11:1.0331967e+00 12:1.0866436e+00 13:1.1237987e+00 14:1.1620448e+00 15:1.1498872e+00 16:1.1502055e+00 17:1.1643373e+00 18:1.1744245e+00 19:1.2277154e+00 20:1.2208029e
+00 21:1.2346281e+00 22:1.2227990e+00
21 1:1.1148580e-01 2:1.4323304e-01 3:1.9382736e-01 4:2.3322868e-01 5:3.0439549e-01 6:1.4267321e-01 7:4.3187887e-01 8:3.2850228e-01 9:7.4925574e-01 10:7.8743849e-01
11:1.0061783e+00 12:9.7755543e-01 13:1.0408131e+00 14:1.0067719e+00 15:9.5916914e-01 16:1.1158977e+00 17:1.0509145e+00 18:1.0701760e+00 19:1.0027518e+00 20:1.1524968e
+00 21:1.0855609e+00 22:9.9566016e-01
21 1:1.6326869e-01 2:1.9830739e-01 3:1.6685063e-01 4:3.3982902e-01 5:1.7169710e-01 6:6.3265527e-01 7:7.5759903e-01 8:1.1226691e+00 9:9.8175718e-01 10:1.1091003e+00
11:1.1156836e+00 12:1.0871267e+00 13:1.0568146e+00 14:1.1383774e+00 15:1.1736896e+00 16:1.1774856e+00 17:1.1588990e+00 18:1.1958570e+00 19:1.3034897e+00 20:1.2036113e
+00 21:1.2317951e+00 22:1.2880625e+00
21 1:2.6886265e-01 2:3.1762120e-01 3:3.9489355e-01 4:3.4080566e-01 5:6.8958532e-01 6:7.3188936e-01 7:8.8911985e-01 8:1.0078712e+00 9:9.6105290e-01 10:1.0735845e+00
11:1.1188973e+00 12:1.0986340e+00 13:1.0633176e+00 14:1.1608539e+00 15:1.1784389e+00 16:1.1807669e+00 17:1.1670611e+00 18:1.1883699e+00 19:1.2367607e+00 20:1.1489216e
+00 21:1.2190462e+00 22:1.1222670e+00
21 1:1.0554266e-01 2:2.3631368e-01 3:2.7213747e-01 4:3.2565270e-01 5:3.5299982e-01 6:4.8977734e-01 7:2.3319595e-01 8:8.8537553e-01 9:9.8337168e-01 10:1.0212980e+00
11:1.0567530e+00 12:9.5723004e-01 13:1.2011440e+00 14:1.0430321e+00 15:1.1067336e+00 16:1.1402162e+00 17:1.1525526e+00 18:1.1476690e+00 19:1.1431238e+00 20:1.1734158e
+00 21:1.1581678e+00 22:1.2004757e+00

24 1:3.3712216e-01 2:4.2008290e-01 3:2.4272443e-01 4:3.9262571e-01 5:2.9394468e-01 6:4.0049682e-01 7:4.2243261e-01 8:4.4409410e-01 9:4.1828196e-01 10:4.7895030e-01
11:3.9482590e-01 12:5.1791616e-01 13:5.6642151e-01 14:4.3687765e-01 15:4.9189774e-01 16:5.0167504e-01 17:5.5735443e-01 18:6.8372607e-01 19:5.2128695e-01
20:5.6408712e-01 21:6.1631143e-01 22:5.4552095e-01
```

FIGURE 3.8 Sample of Selected Speaker's GFCCS and corresponding speaker ID.

The generated feature file by "Feat2SVM" module is then passed through the scaling operation module because the high numeric values of feature parameters may dominate smaller numerical feature values. In testing the scaling range from (−1, 1) to (7, 7) over GFCC values and found that (−5, 5) generates better result in this case. The scaling range is kept the same for training as well as testing phases of SVM classifier. Figure 3.9 represents the sample of scaling operation output over train file.

```
4 1:-3.0419 2:-3.79544 3:-2.18658 4:-2.50851 5:-2.361 6:-2.60851 7:-2.39771 8:-2.32033 9:-1.78118 10:-2.30261 11:-2.76202 12:-2.87136 13:-2.54941 14:-2.56359
15:-2.69495 16:-2.11816 17:-2.39489 18:-2.60944 19:-2.53852 20:-2.10182 21:-2.51176 22:-2.0694
4 1:-3.68994 2:-4.08994 3:-3.79345 4:-4.65305 5:-2.9122 6:-3.08775 7:-2.3712 8:-2.34352 9:-2.06468 10:-1.63972 11:-2.23603 12:-2.22985 13:-2.45734 14:-1.99427
15:-1.8849 16:-1.97465 17:-2.22532 18:-1.9951 19:-2.50985 20:-1.72476 21:-1.81062 22:-1.94506
17 1:1.81236 2:2.96153 3:3.34343 4:2.92719 5:0.629663 6:0.376642 7:1.75919 8:0.76163 9:0.709142 10:1.85395 11:0.508395 12:1.01153 13:-1.62769 14:0.183444 15:0.16129
16:2.02223 17:2.40183 18:1.64065 19:2.05611 20:2.00712 21:0.94236 22:0.241621
```

FIGURE 3.9 Sample of scaling values of train data.

The scaled training file is then exposed to the model generation phase of SVM classifier. Figure 3.10 represents a sample of trained model parameters of SVM classifier.

```
nu = 0.395600
obj = -2.141791, rho = -0.144200
nSV = 7, nBSV = 1
*.*
optimization finished, #iter = 13
nu = 0.382540
obj = -2.198260, rho = 0.130199
nSV = 7, nBSV = 1
.*.*
optimization finished, #iter = 23
nu = 0.450798
obj = -2.470717, rho = -0.043790
nSV = 9, nBSV = 2
*
optimization finished, #iter = 5
nu = 0.909091
obj = -7.634153, rho = -0.676505
nSV = 10, nBSV = 10
Total nSV = 139
basanta@basanta-laptop:~/asr-spkid/svm$
```

FIGURE 3.10 Sample of model parameters of SVM classifier.

The RBF is used as kernel function in SVM for the generation of model parameters and identification of the speaker. Figure 3.11 describes the types of Kernel functions and their parameters.

RBF function is popularly used as kernel functions for a variety of applications of SVM and is defined as:

$$K\left(x, x_{i,}\right) = \text{expi}(x\gamma \,||X - X_i)\,||^2\,) \tag{3.22}$$

```
svm_type c_svc
kernel_type rbf
gamma 0.0454545
nr_class 31
total_sv 139
rho 0.602242 0.0556919 0.43929 0.400559 0.794458 -0.638398 0.487962 0.322967 0.368299 0.0351498 0.385138 0.44777 0.299327 -0.0691588 -0.759332 -0.0818205 0.36281
0.132529 -0.111173 0.0375563 -0.0967749 0.327003 -0.0324729 -0.0684527 0.0870078 -0.0218852 0.0993197 -0.071552 0.639272 0.0418705 -0.735181 -0.346069 -0.248216
-0.410246 -0.838923 -0.261093 -0.445681 -0.460887 -0.570249 -0.313222 -0.150499 -0.421584 -0.645538 -0.7613 -0.706679 -0.494476 -0.578996 -0.578265 -0.581892
-0.592211 -0.570374 -0.541839 -0.599712 -0.494077 -0.595429 -0.497787 -0.534526 -0.57684 -0.650369 0.470076 0.41784 0.389251 -0.58973 0.532042 0.303956 0.372022
0.546166 0.396246 0.48117 0.328108 0.32317 -0.14069 -0.111836 0.364111 -0.0754463 0.479554 -0.0742525 0.18586 -0.0146233 0.386854 0.0676331 0.169146 0.605772 0.669313
0.442481 -0.00465805 -0.143029 -0.0486947 -0.22445 -0.735057 0.0468018 -0.277746 -0.286207 -0.42768 -0.0961819 0.012051 -0.295482 -0.491844 -0.493303 -0.497607
-0.330594 -0.439612 -0.45694 -0.436038 -0.448153 -0.407731 -0.422133 -0.459972 -0.431589 -0.398658 -0.360509 -0.463847 -0.409903 -0.500482 -0.173961 -0.709003
0.113231 -0.0969226 -0.089759 -0.384012 -0.0289779 0.0599937 -0.252018 -0.45543 -0.456932 -0.461414 -0.0480154 -0.401017 -0.418984 -0.397209 -0.409757 -0.367882
-0.382771 -0.421977 -0.392525 -0.353513 -0.31893 -0.426152 -0.369976 -0.464392 -0.979906 0.288655 0.0853776 0.136182 -0.248453 0.155564 0.235498 -0.173399 -0.320725
-0.897785 -0.371987 0.130032 -0.791556 -0.370516 -0.753976 -0.337117 -0.528229 -0.286018 -0.0971592 -0.265988 -0.279687 -0.201417 -0.339363 -0.447361 -0.486062
0.766014 0.651989 0.68601 0.459991 0.698107 0.740582 0.874868 0.26361 0.755979 0.373346 0.682057 0.556335 0.677453 0.775251 0.775819 0.719873 0.559017 0.73526
0.567391 0.496521 0.541305 0.460514 0.716458 0.589824 -0.25577 -0.251849 -0.48185 -0.153435 -0.0519588 -0.351112 -0.537453 -0.53887 -0.542916 -0.261477 -0.48848
```

FIGURE 3.11 Sample of RBF kernel representation of feature parameters.

In the final stage of the proposed voice-based biometric metric system identifies the speaker by comparing the scaled version of test voice password with the trained model of SVM in feature space. Figure 3.12 indicates the sample of scaled test utterance of voice password.

```
test.scale
26 1:-4.27582 2:-4.57864 3:-4.37438 4:-4.57386 5:-4.19915 6:-4.27583 7:-3.32612 8:-2.71886 9:-2.06454 10:-2.11913 11:-2.52199 12:-3.19785 13:-3.11104 14:-3.50366
15:-3.39699 16:-3.64416 17:-3.61457 18:-3.7202 19:-3.35072 20:-3.02674 21:-3.35389 22:-2.67234
20 1:-4.16182 2:-3.95157 3:-3.75953 4:-3.53647 5:-3.61765 6:-2.90854 7:-2.84637 8:-2.34678 9:-2.29263 10:-2.91555 11:-3.18052 12:-3.24189 13:-3.77359 14:-3.79612
15:-3.44131 16:-4.06501 17:-3.66791 18:-4.0961 19:-3.90027 20:-3.42126 21:-3.57016 22:-3.58083
23 1:-3.66135 2:-3.92962 3:-4.0192 4:-3.17295 5:-2.46343 6:-2.71302 7:-2.08803 8:-2.04165 9:-1.38728 10:-1.55598 11:-1.65731 12:-1.67732 13:-1.99914 14:-1.55332
15:-2.00433 16:-1.86412 17:-1.97958 18:-2.24944 19:-2.44175 20:-1.68166 21:-2.00433 22:-1.96363
```

FIGURE 3.12 Sample of scaled version of test voice password.

The average accuracy rate of the voice-based biometric system for identification speaker by using a speech recognition engine as preprocessing module is found as 80%. Figure 3.13 represents the command issued for prediction of the speaker using scaled training samples and the result of the speaker is stored in a file called "res.output."

```
basanta@basanta-laptop:~/asr-spkid/svm$ ./svm-predict test.scale train.scale.model res.output
```

FIGURE 3.13 Prediction of speaker command of speech-based biometric system.

The File, "res.output" retains the information about the true positive and false positive value of voice-based biometric system. Left side entity of arrow represents true speaker ID and right-side entity of arrow represents the predicted speaker ID. Figure 3.14 illustrates predicted speaker ID.

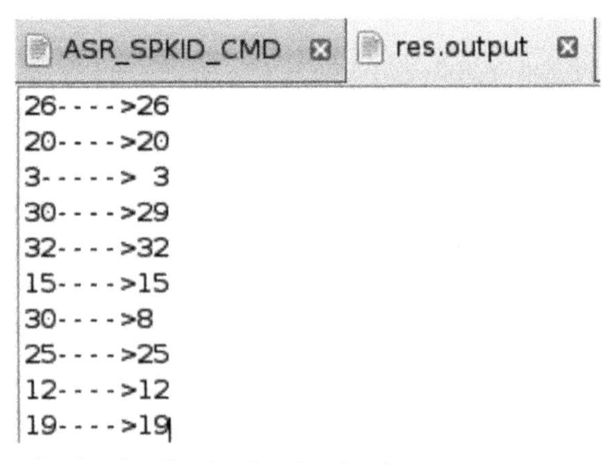
ASR_SPKID_CMD | res.output

```
26---->26
20---->20
3-----> 3
30---->29
32---->32
15---->15
30---->8
25---->25
12---->12
19---->19
```

FIGURE 3.14 Sample of predicted speaker ID.

This chapter has measured the performance rate of the biometric system by varying the scaling ranges of training and testing voice samples. The accuracy rate of the developed system is represented in Figure 3.15 over different scaling factors. It is seen that accuracy rate is reduced drastically by setting the scaling range too low and high and, moreover, a segmentation fault is also detected by the SVM classifier for setting the range very high or low.

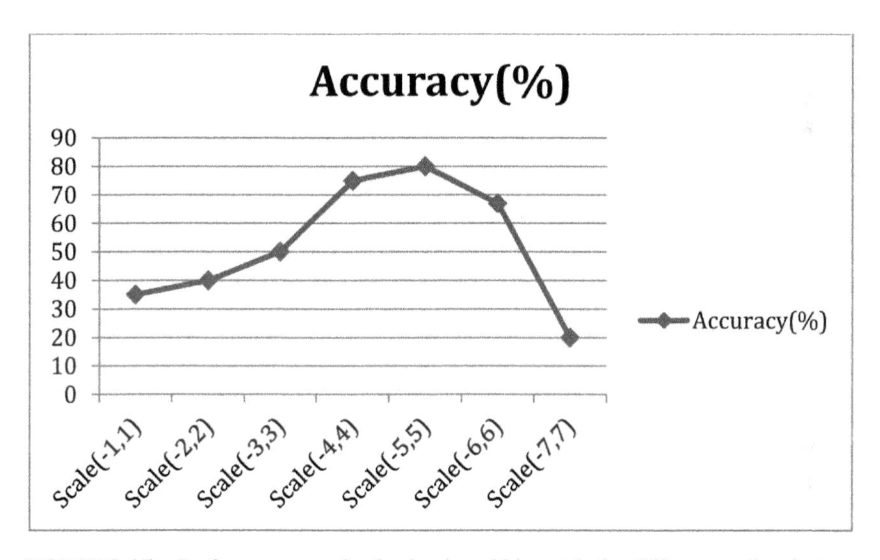

FIGURE 3.15 Performance graph of voice-based biometric for different scaling factors.

3.6 CONCLUSIONS AND FUTURE WORK

In this chapter, a speech-based biometric system is developed for identification speaker using a statistical classifier called SVM. Odia phonetic-based speech recognition engine is used as preprocessing block of biometric system for two key reasons: to enhance the performance rate as well as generate the training model in the fastest manner of a biometric system. The recognition engine is developed using the HMM algorithm that can handle both isolated and continuous voice passwords. The recognition engine is trained using MFCCs and their variants of length 39 as feature parameters because MFCCs calculations follow the human auditory model. The accuracy rate HMM-based speech recognizer is found as 77.1% and

85.1% at the sentence level and word level, respectively. On other hand, SVM classier based on RBF kernel is trained using GFCC. The GFCC is adapted as feature parameters for training and testing scenarios of SVM classifier due to noise robustness property of GFCC. The accuracy rate of speech-based biometric using Odia phonetics is represented in terms of a true positive and false positive. The average accuracy rate of speech based biometric for speaker identification is found as 80%. It is found from the performance curve of the biometric system over different scaling range that the optimal result is obtained at scaling range, (−5,5). Moreover, it is also seen that setting the scaling factor too high and too low leads to a segmentation fault. In this chapter, this is not able to handle imposter due to unavailability mimicry speakers. Hence, future research may be directed on imposter handling. More emphasis is required on incorporation of language identification model in speaker identification task to make the voice-based biometric more universal.

KEYWORDS

- **speech recognition**
- **speaker identification**
- **HMM**
- **SVM**
- **MFCC**
- **GFCC**
- **language model**

REFERENCES

1. O'Shaughnessy, D. *Speech Communications: Human and Machine*, 2nd ed; Wiley-IEEE Press: Hoboken, New Jersey, USA, 2007.
2. Huang, X.; Acero, A.; Hon, H. *Spoken Language Processing: A Guide to Theory, Algorithm and System Development*, Prentice Hall: Upper Saddle River, New Jersey, USA.

3. Jurafsky, D.; Martin, J. H. *Speech and Language Processing: An Introduction to Natural Language Processing, Computational Linguistics, and Speech Recognition*, Pearson Education Asia: Upper Saddle River, New Jersey, USA.
4. Minker, W.; Bennacef, S. *Speech and Human-Machine Dialog*, Kluwer Academic Publishers: Boston, USA.
5. Furui, S. 50 years of Progress in Speech and Speaker Recognition Research. *J. Acoustical Soc. Am.* **2004**.
6. Kinnunen, T.; Li, H., An Overview of Text-independent Speaker Recognition: From Features to Supervectors. Speech Commun. **2010,** *52* (1), 12–40.
7. Lei, Y.; Hansen, J. *Speaker Recognition Using Supervised Probabilistic Principal Component Analysis*, In Proc. Interspeech, 2010, 382–385.
8. Praat software website: http://www.fon.hum.uva.nl/praat. (accessed May 14, 2018)
9. Nakagawa, S.; Wang, L.; Ohtsuka, S. Speaker Identification and Verification by Combining MFCC and Phase Information. *IEEE Transactions on Audio, Speech, and Language Processing* **2012,** *20* (4), 1085–1095.
10. Lu, X.; Dang, J. An Investigation of Dependencies Between Frequency Components and Speaker Characteristics for Text-independent Speaker Identification. *Speech Commun.* **2008,** *50* (4), 312–322.
11. Rabiner, L. R. In *A Tutorial on Hidden Markov Models and Selected Applications in Speech Recognition*. Proceedings of the IEEE, 1989, 256–286.
12. Wachter, M. D.; Matton, M.; Demuynck, K.; Wambacq, P.; Cools, R.; Compernolle, D. V. Template-Based Continuous Speech Recognition. *IEEE Transactions on Audio, Speech and Language Processing* **2007,** *15* (4), 1377–1390.
13. Wang, D. L.; Brown, G. J. *Computational Auditory Scene Analysis: Principles, Algorithms, and Applications;* Wiley-IEEE Press: Hoboken, NJ, 2006.
14. Tazi, E. B.; Benabbou, A..; Harti, M. Efficient Text-Independent Speaker Identification Based on GFCC and CMN Methods. *ICMCS* **2012,** 90–95.
15. Manning, C. D.; Schutze, H. *Foundations of Statistical Natural Language Processing;* 2nd ed; MIT Press: USA, 2000.
16. Rabiner, L. R.; Schafer, R. W. *Digital Processing of Speech Signals*, Pearson Education, 1st ed; 2004.
17. Quatieri, T. F. *Discrete-Time Speech Signal Processing Principles and Practice*, Pearson Education, Third Impression, 2007.
18. Rose, P. *Forensic Speaker Recognition*, Taylor and Francis, Inc.: New York.
19. Alam, M. J.; Kenny, P.; O'Shaughnessy, P. *Robust Feature Extraction for Speech Recognition by Enhancing Auditory Spectrum*. Proc. INTERSPEECH: Portland Oregon, 2012.
20. Mohanty, S; Swain, B. K. *Double Ended Speech Enabled System in Indian travel & Tourism industry*. In proceedings of 2013 IEEE International Conference on Computational Intelligence and Computing Research, Tamil Nadu, India, 2013.
21. Sahidullah, M. Saha, G. *Design, Analysis and Experimental Evaluation of Block Based Transformation in MFCC Computation for Speaker Recognition. Speech Commun.* **2012,** *54* (4), 543–565.
22. Kim, C.; Stern, R. M. *Feature Extraction for Robust Speech Recognition Using a Power-law Nonlinearity and Power-bias Subtraction*, Proc. INTERSPEECH, Sept 2009.

23. Patterson, R. D.; Nimmo-Smith, I.; Holdsworth, J.; Rice, P. An Efficient Auditory Filterbank Based on the Gammatone Function. Appl. Psychol. Unit, Cambridge University.
24. Samudravijaya K.; Barot. M. *A Comparison of Public Domain Software Tools for Speech Recognition*, Workshop on Spoken Language Processing, 2003, pp 125–131.
25. http://www.speech.cs.cmu.edu/sphinx/tutorial.html. (accessed May 14, 2018)
26. Becchetti, C.; Ricotti, L. P. *Speech Recognition Theory and C++ Implementation;* Jhon Wiley & Sons: London, UK, 2009.
27. Polur, P. D.; Miller, G. E. Experiments with Fast Fourier Transform, Linear Predictive and Cepstral Coefficients in Dysarthric Speech Recognition Algorithms Using Hidden Markov Model. *IEEE Trans. Neural Syst. Rehabil. Eng.* **2005,** *13* (4), 558–561.
28. Keshet, J.; Bengio, S. *Automatic Speech and Speaker Recognition: Large Margin and Kernel Methods;* John Wiley & Sons.
29. Gales, M.; Young, S. The Application of Hidden Markov Models in Speech Recognition. *Foundations and Trends® in Signal Processing;* Now Publishers Inc.: Hanover, MA, USA, 2008; Vol. 1 (3), pp 195–304.
30. Duda, R. O., Hart, P. E.; Stork, D. G. *Pattern Classification,* 2nd ed.; Wiley-Interscience New York: NY, USA, 2000.
31. Bishop, C. M. *Pattern Recognition and Machine Learning;* Springer: Cambridge CB3 0FB, U.K, 2006.
32. Vapnik, V. *The Nature of Statistical Learning Theory;* Springer-Verlag: New Jersey, USA, 1995.
33. Mohanty, S.; Swain, B. K. *Identification of Odia Dialects using Kernels of SVM*, In Proceedings of 23rd International Symposium Frontiers of Research in Speech and Music (FRSM), NIT, Rourkela, India, 2017.
34. Keshet, J.; Bengio, S. *Automatic Speech and Speaker Recognition: Large Margin and Ker n el Methods;* John Wiley & Sons Ltd.: West Sussex, United Kingdom, 2009.
35. Schlkopf, B.; Smola, A. J. *Learning with Kernels: Support Vector Machines, Regularization, Optimization, and Beyond Adaptive Computation and Machine learning;* MIT Press: Cambridge.
36. James, G., Witten, D.; Hastie, T.; Tibshirani, R. *An Introduction to Statistical Learning with Applications in R;* Springer: New York Heidelberg Dordrecht London, 2013.
37. Campbell, W. M.; Sturim, D. E.; Reynolds, D. A. Support Vector Machines Using GMM Supervectors for Speaker Verification. *IEEE Signal Process. Lett.* **2006,** *13* (5), 308–311.
38. Chang, C.; Lin, C. -J. LIBSVM: A Library for Support Vector Machines. *ACM Trans. Intell. Syst. Technol.* **2011,** *2* (3), *27*, 1–27.
39. Chowdhary, C. L. *Linear Feature Extraction Techniques for Object Recognition: Study of PCA and ICA*, *J. Serbian Soc. Comput. Mech.* **2011,** *5* (1), 19–26.
40. Chowdhary, C. L.; Acharjya, D. P. Singular Value Decomposition–Principal Component Analysis-Based Object Recognition Approach. *Bio-Inspired Computing for Image and Video Processing*, 2018, 323.

CHAPTER 4

Deep Learning Techniques to Classify and Analyze Medical Imaging Data

DHARM SINGH JAT and CHUCKNORRIS GARIKAYI MADAMOMBE*

Department of Computer Science, Namibia University of Science and Technology, Windhoek, Namibia

**Corresponding author. E-mail: chuckygari@gmail.com*

ABSTRACT

The use of deep learning techniques predominantly the Convolutional Neural Networks (CNNs) has been used in various disciplines in recent years. CNNs have shown an essential ability to automatically extract large volumes of information from big data. The use of CNNs has significantly proved to be useful especially in classifying natural images. Nonetheless, there has been a major barrier in implementing CNNs in the medical domain due to lack of proper available data for training. As a result, generic imaging standards like ImageNet have been widely used in the medical domain. However, these generic imaging benchmarks are not so perfect as compared to the use of CNNs. The main aim of this chapter is review the existing deep learning techniques to classify and analyze medical imaging data. In this chapter, a comparative analysis of LeNet, AlexNet, and GoogLeNet was done. Furthermore, this review looked at the literature on classifying medicinal anatomy images using CNNs. Based on the various studies the CNNs architecture has better performance with the other architectures in classifying medical images.

4.1 INTRODUCTION

In the United States, millions of medical images are taken every day from both private and state hospitals (Litjens et al., 2017). The same applies to all other countries throughout the world. There is a lot of images that are taken in hospitals on a daily basis (Litjens et al., 2017). As a result, there is a considerable increase in pressure on the healthcare providers for them to provide accurate and efficient diagnostic services.

According to Zhao and Xie (2012), the National Institute of Medicine has estimated that approximately more that 15 million patients in America are wrongly diagnosed on a yearly basis. This is quite a large number of people who are being misdiagnosed and it is a major problem if this continues. This problem is emanating from the fact that there are large volumes of imaging data produced on a daily basis in the hospitals. In addition, there are no proper systems in place that can be used to accurately and efficiently handle such kind of big data (Litjens et al., 2017). As a result, there is a need for more precise and resourceful decision support systems for doctors so that they can significantly reduce the number of patients who are misdiagnosed (Rajchl et al., 2017).

According to Li et al. (2017), deep learning technology works in a similar way as the working of the human brains. The artificial neuron networks have the ability to examine enormous amounts of data in order to automatically discern essential patterns without the need for humans to be available (Gal and Ghahramani, 2016). In fact, the artificial neuron networks can work much better as compared to humans. In other words, artificial neural networks work in a similar way as humans do. However, since they are systems, artificial neural networks perform better as compared to human beings. In most cases, deep learning is mainly used where there is a need for classifying patterns in amorphous data (Kamnitsas et al., 2017). Thus deep learning is mostly used to identify unstructured patterns in various media such as text, sound, video, and medical images (Wang et al., 2018).

When the doctors are examining the patients, they try always to solve sophisticated problems. In most cases, the goal of many medical doctors is to give the correct treatment to the patients based on the available previous medical reports like the lab test reports, signs and symptoms of the patients, medical images as well as the patient's medical history (Singh and Singh, 2018). A study conducted by Yigzaw and Bellika (2014) indicated that digital healthcare data is estimated to grow sharply from 500 petabytes to 25,000 petabytes by 2020 throughout the whole world. As noted earlier, it is a big challenge for medical doctors to get precise understandings from billions of clinical data. As a result, a large number of researchers, medical professionals, and data scientists are continuously finding solutions to advance patient care in clinics and hospitals.

In this modern-day of improved technology, there is a need to implement deep learning in the medical industry. According to Wang et al. (2018), machine learning algorithms are able to do information processing and pattern recognition and identification in a better way as compared to human beings. In addition, machine learning algorithms can be used to comprehend risk factors for diseases in a very large population. Furthermore, machine learning algorithms can also be used to identify and predict dangerous diseases such as cancer, diabetes, etc.

According to Singh and Singh (2018), the use of computer-assisted diagnosis (CAD) to assess scans of women can detect approximately 52% of cancer before the women were diagnosed officially. Ker et al. (2017) noted that machine learning algorithms can be used in various disciplines of medicine including the discovery of drugs and decision-making in clinical. In addition, the use of machine learning algorithms can change by a huge margin the way in which medicine is practiced to date. The power of machine learning algorithms in recent years has come at a time when the medical records are being digitized. Unlike in the past, when medical records were mainly paper-based, these days, most medical records are being stored electronically. Machine learning algorithms cannot work with paper-based medical records. They can only work if the medical records are digitized. This means that these machine learning algorithms have come at the right time when the medical records are now being digitized.

According to Ker et al. (2017), the use of electronic health records (EHR) in recent years has increased sharply from approximately 12% to 40% in the United States from 2007 to 2012. Despite the fact that medical images are an important component of any patient's EHR, they are currently being analyzed manually by human radiologists (Hsiao et al., 2014). Human beings cannot be compared to machines because they are slower, they get tired and they might not have much experience. All these are the major limitations of using humans as compared to machine learning algorithms. According to Hsiao et al. (2014), a diagnosis that is delayed and incorrect can be fatal to patients. As a result, it is crucial to automate medical image analysis through the use of precise and effective machine learning algorithms.

On a daily basis, there is an increase in the number of medical images, for instance, CT, MRI, and X-ray. These types of medical images which are increasing on a daily basis are crucial because they provide vital information in order for doctors to provide an accurate diagnosis, medical treatments, education as well as providing medical research (Müller et al., 2004). In general, the usual methods used to retrieve medical images rely on the annotation of keywords. However, relying on images annotation is not efficient because the process takes a lot of time and also it is difficult to describe the contents of these images with words (Qiu et al., 2017).

In recent years, the content-based image retrieval (CBIR) has been used more often in the medical image retrieval as well as in the classification of images. The increase in the use CBIR is due to its improved computing power as well as the considerable gain and popularity in the applications of medical image retrieval and classification. It is noted that the dawn of deep learning techniques has conveyed a complete change especially in the field of machine learning. The deep learning techniques and the machine learning algorithms have shown better performances as compared to the use of traditional approaches. Other than its use in the classification, the other powerful aspect of deep learning is that they do not require any designing features. Using deep learning, these features can be learned from the raw data which is not processed yet.

According to LeCun et al. (2017), deep learning is a rapidly emergent technology that is mainly focused on data analysis. Deep learning has been regarded as one of the top technologies in 2013 in analyzing medical images. Generally, deep learning can be regarded as an enhancement of artificial neural networks. Deep learning differs in artificial neural networks in that it consists of increased layers that allow higher levels of abstraction and enhanced estimates from the data (LeCun et al., 2017). In this modern day and age, it appears that CNN is the top machine learning tool in the computer vision as well as the general medical imaging domains. CNNs have proven to be very powerful especially in classifying and analysing medical imaging.

4.2 TYPES OF MEDICAL IMAGING

In recent years, there is a tremendous increase in the use of images in the healthcare environment. According to Zhao and Xie (2012), the different types of digital medical images that exist are an ultrasound (US), X-ray, computed tomography (CT) scans and magnetic resonance imaging (MRI) scans, positron emission tomography (PET) scans, retinal photography, histology slides, and dermoscopy images. In general, CT and MRI scans examine multiple organs while retinal photography and dermoscopy examine specific organs. Zhao and Xie (2012) conducted a study on imaging use for a period from 1996 to 2010 across six large integrated healthcare systems in the United States. The authors concluded in their study that CT, MRI, and PET usage increased by 7.8%, 10%, and 57%, respectively.

4.3 THE NEED FOR CNN IN MEDICAL DOMAIN

In recent years, there have been a lot of images that are taken in hospitals on a daily basis (Litjens et al., 2017). This, in turn, has increased pressure on the health care for them to provide accurate and efficient diagnostic services. The National Institute of Medicine has estimated that approximately more than 15 million patients in America are wrongly diagnosed on a yearly basis (Zhao and Xie, 2012). This is quite a huge number of people who are being misdiagnosed and it is a major problem if this continues. This problem is emanating from the fact that there are large volumes of imaging data produced on a daily basis in the hospitals and there is no proper systems in place that can be used to accurately and efficiently handle such kind of data (Litjens et al., 2017). As a result, there is a need for more precise and resourceful decision support systems for doctors so that they can significantly reduce the number of patients who are misdiagnosed (Rajchl et al., 2017).

The human brain can recognize an object much faster but for a computer system accuracy for an object recognition depends on the use of the tools and the level of algorithms. Chowdhary et al. (2015) present literature of a three-dimensional object recognition system and connecting such algorithm with 3D object recognition intelligence system to identify individuals. Chowdhary et al. (2018) study focuses on detecting outliers, rare events, anomalies, vicious actions, exceptional phenomena that occur in database.

Over the past decades, the use of medical imaging worldwide has been increasing sharply. The most common medical imaging that has been increasing in the past decade are: computed tomography (CT), magnetic resonance imaging (MRI), positron emission tomography (PET), mammography, ultrasound, and X-ray. In most cases, these medical images are analyzed by a human expert. However, the examination of medical images by a human expert is sometimes time consuming and it is subject to human errors. Therefore, manual examination of medical imaging by human experts is sometimes not accurate. The use of deep learning in recent years has provided improved solutions for medical image analysis in the health industry. The coming up of deep learning techniques has helped medical doctors to identify, classify, quantify and analyze patterns in medical images. The proposal has outlined the applicability and the major advantages of deep learning concepts, particularly the CNN in medical imaging analysis.

According to Ker et al. (2017), the use of electronic health records (EHR) in recent years has increased sharply from approximately 12–40% in the United States from 2007 to 2012. Despite the fact that medical images are an important component of any patient's EHR, they are currently being analyzed manually by human radiologists (Hsiao et al., 2014). Human beings cannot be compared to machines because they are slower, they get tired and they might not have much experience. All these are the major limitations of using humans as compared to machine learning algorithms. According to Hsiao et al. (2014), a diagnosis that is delayed and incorrect can be fatal to patients. As a result, it is crucial to automate medical image analysis through the use of precise and effective machine learning algorithms. The main aim of this chapter is to review the existing deep learning techniques to classify and analyze medical imaging data.

In this modern day of improved technology, there is a need to implement deep learning in the medical industry. According to Wang et al. (2018), machine learning algorithms are able to do information processing and pattern recognition and identification in a better way as compared to human beings. In addition, machine learning algorithms can be used to comprehend risk factors for diseases in a very large population. Furthermore, machine learning algorithms can also be used to identify and predict dangerous diseases such as cancer, diabetes, etc.

According to Singh and Singh (2018), the use of computer-assisted diagnosis (CAD) to assess scans of women can detect approximately 52% of cancer before the women were diagnosed officially. Ker et al. (2017) noted that machine learning algorithms can be used in various disciplines of medicine including the discovery of drugs, decision-making in clinical. In addition, the use of machine learning algorithms can change by a huge margin the way in which medicine is practiced to date. The power of machine learning algorithms in recent years has come at a time when the medical records are being digitized. Unlike in the past when medical records were mainly paper-based, these days, most medical records are being stored electronically. Machine learning algorithms cannot work with paper based medical records. They can only work if the medical records are digitized. This means that these machine learning algorithms has come at the right time when the medical records are now being digitized.

Dhara et al. (2016) conducted a study on the grouping of anatomies in the lung images. The study concluded that the use of deep learning techniques has shown good results in the classification and analyzing of images. LeCun et al. (2016) examined the use of CNNs for digit recognition. The use of CNN proved to be very successful based on the positive results that they got. Krizhevsky et al. (2012) also proposed a deep CNN architecture that turned out to be very successful.

4.4 TECHNICAL APPROACH

These days the Python programming language are using for machine learning instead of other programming languages because of the following reasons:

1. Python has a wide range of packages for Machine Learning, for example, numpy, pandas, keras, and tensorflow.
2. Python libraries are powerful to perform complex computations.
3. Python programming language is user-friendly as well as easy to learn.

4.4.1 *MACHINE LEARNING PROCESS*

Figure 4.1 below illustrates the overall process of machine learning in medical imaging (Dhara et al., 2016).

1. *Get data*—this is a set of data that is used to train the machine learning algorithm. In this case, this will be a set of medical imaging data.
2. *Store data*—data is stored in a database for easy retrieval purposes.
3. *Load and analyze data*—the data will be loaded from the datasets and basic data analysis will be performed.
4. *Transform data*—only numeric data is required for machine learning. Therefore, there is a need to transform the input data.
5. *Learn (fit)*—run the labeled data through a machine learning algorithm yielding a model.

6. *Predict*—use the model to predict labels for data that the model did not see previously.
7. *Assess*—verify the accuracy of predictions made by the model.

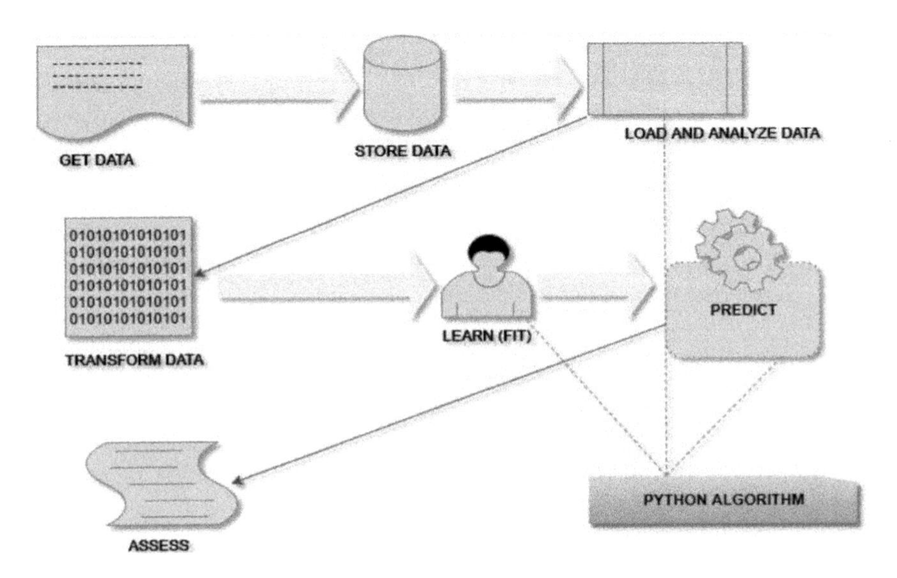

FIGURE 4.1 Machine learning overview.
Source: Dhara et al. (2016).

4.4.2 THE CNN ARCHITECTURE

Khan and Yong (2017) proposed CNN architecture which is an improvement of the AlexNet architecture. Further this architecture was made up of four convolutional layers.

In AlexNet architecture, the convolutional layers were five and the three fully connected layers. In the proposed architecture, only four convolutional layers will be used. The outputs from the convolution layers will be calculated as follows:

$$Y^{\iota}_{I,j} = \sum W_{ab} * X^{\iota-1}_{(i+a)(j+b)} \quad (4.1)$$

Equation 4.2 below shows the features that map the results from convolution architecture.

$$Y_{ij} = \max \{0, Y_{IJ}) \quad (4.2)$$

4.5 ARTIFICIAL INTELLIGENCE-RELATED ANALYSIS BASED ON CNN

The proposed CNN architecture was evaluated according to the accuracy as compared to the other conventional methods for medical imaging classification and analysis. In this study, a comparative analysis of LeNet, AlexNet, and GoogLeNet was done. Thereafter, the study reviewed an improved conceptual framework for classifying medicinal anatomy images CNNs. The contribution of this chapter is on a comprehensive review of the three milestone CNN architectures, that is, LeNet, AlexNet, and GoogLeNet for classifying medical anatomy images. The findings from the performance analysis of these architectures advocates the need of a modified architecture because of their poor performance for medical image anatomy classification.

Table 4.1 below has examined some major applications of CNN for computer-aided diagnosis. The three major diagnosis issues illustrated by Table 4.1 below are thyroid nodule diagnosis, breast cancer diagnosis, and diabetic retinopathy. Furthermore, the table has illustrated the various dataset that has been used for each application.

TABLE 4.1 CNN for Computer Aided Diagnosis.

Author	Application	Method	Dataset	Accuracy
Valerio et al. (2019)	Breast lesions	Pipeline for effective classification of mammograms	Used the public dataset called MAMMOSET	~94.34%
Okamoto et al. (2019)	Narrow-Band Imaging (NBI) colorectal endoscopic images	Pre-learned CNN as a feature extraction module	NBI image dataset including 1260 NBI patches collected from endoscopic examinations at Hiroshima University Hospital	~90%
Ma et al. (2017)	Diagnosis of the thyroid nodule	The use of pre-trained CNN	Used the ultrasound images	~83%
Khan and Yong (2017)	Medical image anatomy classification	Four convolutional layers followed by two fully connected layers	Acquired from the U.S. national library of medicine	81%

TABLE 4.1 *(Continued)*

Author	Application	Method	Dataset	Accuracy
Sun et al. (2016)	Diagnosis of Breast Cancer	CNN using semi supervised learning	Used of the Mammographic Images with ROIs	82.43%
Pratt et al. (2016)	The use of Diabetic Retinopathy	CNN	The used of Kaggle Dataset	75%

From Table 4.1, it can be noted that the use of CNN in computer-aided diagnosis is very helpful since the accuracy is higher. However, pre-learned CNN as a feature extraction module will produce more accuracy percentage since it is a hybrid combination of various architectures.

The open-access medical image database that contains millions of anonymous marked medical imaging data was used to do the experiments (Khan and Yong, 2017). Further in this study, the medical imaging modalities that were used in the experiment consisted of five anatomies as follows; CT, MRI, PET, X-RAY, and ultrasound. To evaluate the experiments, approximately 37,198 images of these five anatomies were used to train the CNN models. Approximately 500 images were used for testing purposes. A total of 100 from each anatomy set were used. The medical imaging anatomies that were used in the experiment were: lungs, liver, heart, kidney, and lumbar spine (Khan and Yong, 2017).

Further in this study, the medical imaging modalities. Further, the proposed dataset was compared and tested against the other previous CNN architectures that have been developed by other studies.

The variables that were used to compare the existing architectures and the proposed CNN were runtime, training loss, validation accuracy, and test accuracy.

Deep learning architectures particularly Convolutional Neural Network (CNN) compare with LeNet, AlexNet, and GoogleNet (Khan and Yong, 2017). Proposed CNN has 89% test accuracy which is much higher as compared to the AlexNet which has an accuracy of only 74%, and the LeNet which has an accuracy of 59%. The GoogleLeNet was having the lowest test accuracy of 45%. In the proposed CNN, the validation loss has been reduced to 0.67. Whereas in LeNet, AlexNet, and GoogLeNet the validation loss was 1.3, 1.39, and 1.2, respectively. The validation accuracy for proposed CNN was found to be 76.6%. This was very high as compared to the LeNet, AlexNet, and GoogLeNet which were 58%, 65%, and 55%, respectively. The CNN had a run time of 16,728 seconds which is lower than all the other architectures.

Pre-trained CNN combined with the transfer learning technique is the good choice to classify general image datasets. Valerio et al. (2019) present on the breast lesions which is one of the most common types of cancer suffering women worldwide. Further, in this study to improve the learning of the networks the proposed methodology applies a data augmentation method. This method capable of analyzing different approaches, regarding handcrafted and deep features, through end-to-end networks and traditional classifiers to automatically answer the best tuning according to a given mammographic image dataset. The proposed method also validated on public image datasets and study achieved up to 94.34% classification accuracies.

4.6 CONCLUSIONS

This chapter reviews the existing deep learning techniques to classify and analyze medical imaging data. The chapter discussed the various CNN architecture that consists of numerous intricacy layers. CNN architecture has outclassed the existing main architectures used for classifying medical images in the health sector. However, it has been noted that medical imaging anatomies are different from natural images due to privacy and security policies. Pre-trained convolutional neural networks combined with the transfer learning technique is the good choice to classify general image datasets.

KEYWORDS

- **AlexNet**
- **convolutional neural networks**
- **GoogLeNet**
- **ImageNet**
- **LeNet**

REFERENCES

Chowdhary, C. L.; Muatjitjeja, K.; Jat, D. S. Three-Dimensional Object Recognition Based Intelligence System for Identification, Emerging Trends in Networks and Computer Communications (ETNCC), 2015.

Chowdhary, C. L.; Ranjan, A.; Jat, D. S. Categorical Database Information-Theoretic Approach of Outlier Detection Model. *Ann. Comput. Sci. Ser.* 14th Tome 2nd Fasc. **2016,** *2016*, 29–36.

Dhara, A. K.; Mukhopadhyay, S.; Dutta, A.; Mandeep, G.; Niranjan, K. A Combination of Shape and Texture Features for Classification of Pulmonary Nodules in Lung CT Images. *J. Dig. Imag.* **2016,** *29*, 466–475.

Felipe, J. C.; Traina, A. J. M.; Traina, C.. Retrieval by Content of Medical Images Using Texture for Tissue Identification. *IEEE Symp. Comput. Based Med.* **2017,** *12*, 175–180.

Gal, Y.; Ghahramani, Z. Dropout as a Bayesian Approximation: Representing Model Uncertainty in Deep Learning. *ICML* **2016,** *17*, 1050–1059.

Hsiao, C. J.; Hing, E.; Ashman, J. Trends in Electronic Health Record System Use Among Office-Based Physicians. *Nat. Health Stat. Rep.* **2014,** *75*, pp 1–18.

Kamnitsas, K.; Ledig, C.; Newcombe, V. F. J.; Simpson, J. P.; Kane, A. D.; Menon, D. K.; Rueckert, D.; Glocker, B. Efficient Multi-Scale 3D CNN With Fully Connected CRF for Accurate Brain Lesion Segmentation. *Med. Image Anal.* **2017,** *36*, 61–78.

Ker, J.; Wang, L.; Rao, J.; Lim, T. Deep Learning Applications in Medical Image Analysis. *IEEE Access* **2017,** *6*, 9375–9389.

Khan, S.; Yong, S. P. A Deep Learning Architecture for Classifying Medical Images of Anatomy Object. *2017 Asia-Pacific Signal and Information Processing Association Annual Summit and Conference (APSIPA ASC),* 2017.

Krizhevsky, A.; Ilya, S.; Hinton, G. E. In *Imagenet Classification With Deep Convolutional Neural Networks*. Advances in Neural Information Processing Systems, 2012; pp 1097–1105.

Lecun, Y.; L′Eon, B.; Yoshua, B.; Haffner, P. Gradient-Based Learning Applied to Document Recognition. *Proc. IEEE* **2016,** *86*, 2278–2324.

Lecun, Y.; Bengio, Y.; Hinton, G. Deep Learning Techniques. *Nature* **2017,** *521*, 436–444.

Li, W.; Wang, G.; Fidon, L.; Ourselin, S.; Cardoso, M. J.; Vercauteren, T. On the Compactness, Efficiency, and Representation of 3D Convolutional Networks: Brain Percolations as a Pretext Task. *IPMI* **2017,** *13*, 348–360.

Litjens, G.; Kooi, T.; Bejnordi, B. E.; Setio, A. A. A.; Ciompi, F.; Ghafoorian, M.; Van Der Laak, J. A.; Ginneken, B. V.; S'Anchez, K. A Survey on Deep Learning in Medical Image Analysis. *Med. Image Anal.* **2017,** *42*, 60–88.

Mohammad, R. Z.; Ahmed, M.; Woo, C. S. Automatic Medical X-Ray Image Classification Using Annotation. *J. Dig. Imag.* **2014,** *27*, 77–89.

Müller, H.; Michoux, N.; Bandon, D.; Geissbuhler, A. A. Review of Content Based Image Retrieval Systems in Medical Applications: Clinical Benefits and Future Directions. *Int. J. Med. Inform.* **2004,** *73*, 1–23.

Muller, H.; Rosset, A.; Vallee, J. P.; Geisbuhler, A. Comparing Feature Sets for Content-Based Image Retrieval in a Medical Case Database. *SPIE Med. Image PACS Image* **2004,** 99–109.

Takumi, O.; M, O.; Tetsushi, K.; Shinji, T.; Toru, T.; Bisser, R.; Kazufumi, K.; Shigeto, Y.; Hiroshi, M. Feature Extraction of Colorectal Endoscopic Images for Computer-Aided Diagnosis with CNN, *2019 2nd International Symposium on Devices, Circuits and Systems (ISDCS)*, 2019. DOI: 10.1109/ISDCS.2019.8719104.

Qiu, C.; Cai, Y.; Gao, X.; Cui, Y. Medical Image Retrieval Based on the Deep Convolution Network And Hash Coding. *Int. Congr. Image Signal Process. Biomed. Eng. Inform.* **2017,** *1*, 1–6.

Rajchl, M.; Lee, M.; Oktay, O.; Kamnitsas, K.; Palmbach, J.; Bai, W.; Rutherford, M.; Hajnal, J.; Kainz, B.; Rueckert, D. Object Segmentation From Bounding Box Annotations Using Convolutional Neural Networks. *TMI* **2017,** *36*, 674–683.

Singh, S.; Singh, N. In *Object Classification to Analyze Medical Imaging Data Using Deep Learning*. International Conference on Innovations in Information Embedded and Communication Systems (ICIIECS), 2018; Vol. 1, pp 1 – 4.

Valerio, L. M.; Daniel H. A. Alves, L.; F. Cruz, P. H.; Bugatti, C. O.; Priscila, T. M.; Saito, D. M. Deep Transfer Learning for Lesion Classification of Mammographic Images, *2019 IEEE 32nd International Symposium on Computer-Based Medical Systems (CBMS),* 2019. DOI: 10.1109/CBMS.2019.00093

Wang, G.; Li, W.; Maria, A.; Zuluaga, P. R.; Premal, A.; Aertsen, M.; Doel, T.; Anna. L. J.; Ourselin, D. S.; Vercauteren, T. Interactive Medical Image Segmentation Using Deep Learning With Image-Specific Fine-Tuning. *IEEE Trans. Med. Imag.* **2018,** *37* (7) 1562–1573.

Wei, Y.; Zhentai, L.; Mei, Y.; Meiyan, H.; Qianjin, F.; Wufan, C. Content-Based Retrieval of Focal Liver Lesions Using Bag-Of-Visual-Words Representations of Single-And Multiphase Contrast-Enhanced CT Images. *J. Dig. Imag.* **2012,** *25*, 708–719.

Yigzaw, K. Y.; Bellika, J. G. In *Evaluation of Secure Multi-Party Computation for Reuse of Distributed Electronic Health Data*. IEEE-EMBS International Conference On Biomedical And Health Informatics (BHI), 2014; Vol. 15, pp 219–222.

Zhao, F.; Xie, X. An Overview of Interactive Medical Image Segmentation. *Ann. BMVA* **2012,** *7*, 1–22.

CHAPTER 5

Face Recognition System: An Overview

SHUBHAM SHRIMALI and CHIRANJI LAL CHOWDHARY*

Department of Software and System Engineering,
School of Information Technology and Engineering, VIT Vellore, India

**Corresponding author. E-mail: c.l.chowdhary@gmail.com*

ABSTRACT

Face recognition is the discipline which includes the appreciative of the way the faces are recognized by biological systems and the way it is emulated by computer systems. Genetic systems occupation dissimilar kinds of visual instruments to have been intended by nature to ensemble a certain location where the agent subsists. Also, computer systems occupation unlike visual expedients to imprisonment and development faces as greatest designated in apiece certain claim. Such sensors can be video cameras, ultraviolet cameras, or 3-dimesional scans. Face recognition system is a biometric technology used to identify or verify the user by comparing facial features to database. It is a fast growing, challenging, and interesting area in biometric systems. It is generally used for security purpose that can be compared with other biometrics, like fingerprint, iris, etc.

5.1 INTRODUCTION

Face recognition system is a biometric technology used to identify or verify the user by comparing facial features to database. It is a fast growing, challenging, and interesting area in biometric systems. It generally used for security purpose that can be compared with other biometrics, like fingerprint, iris, etc. From the last 20 years it has become one of the most popular areas of research. As we know user privacy and the data are very important and to protect it, we need a security. There are so many security systems or techniques to secure the data, but every system has some loophole and to overcome from this face recognition is used. In this system there is no need to remember the password; no one can steal your password. The password is your face and no one can access the data without your permission. In 1960, the first face recognition system was created. Since then, face recognition system come a long way. There so many techniques came to improve the system and get more accurate result example eigenface, Gabor filters, support vector machine (SVM), neural networks, 3D recognition, etc. Now with novel techniques and algorithm it is more challenging to find out which is the best technique to use for security. Face recognition is one of the best security systems. In these facial features are extract, like size of nose, eyes, lips, etc. All these features are unique for every person. These features will be stored in database, whenever we want to verify the user; we match these features to database. There are some drawbacks, like in case of twins it cannot differentiate which one is the valid user. Secondly if someone use the photograph of user, then system cannot differentiate whether it is real person or the photograph. In this chapter, a complete survey is done on the basis of possible techniques to overcome the drawbacks of security.

From the last few years biometrics received lots of attention in every sector. It is a best and secure way to identify the user. In this user does not need to remember the password and there is no tension of forgetting the plastic card because this is user-friendly and the biological appearance of human never changes. There are many different biometric technologies, such as finger print recognition, iris, palm, etc., but face recognition gets more attention in computer vision area. In 1950s[1] and 1960s[2] the first face recognition research was conducted. But still face recognition is a difficult task because human face is very complex and challenging to extract the features. Over the past few decades, face recognition has become the most used application in the surveillance and in security. Face recognition is the most challenging aspect in the field of image analysis because every person has a different facial feature. Face recognition has been biased toward the visible spectrum for a variety of reasons. The visible spectrum image is constructed by a reflection on the surfaces.[3–22,152–165,119] One of the major drawbacks is lighting and skin color.[23] There are several methods that are proposed to overcome from the problem of lighting that fall into two main categories: passive methods and active methods (Fig. 5.1).[24]

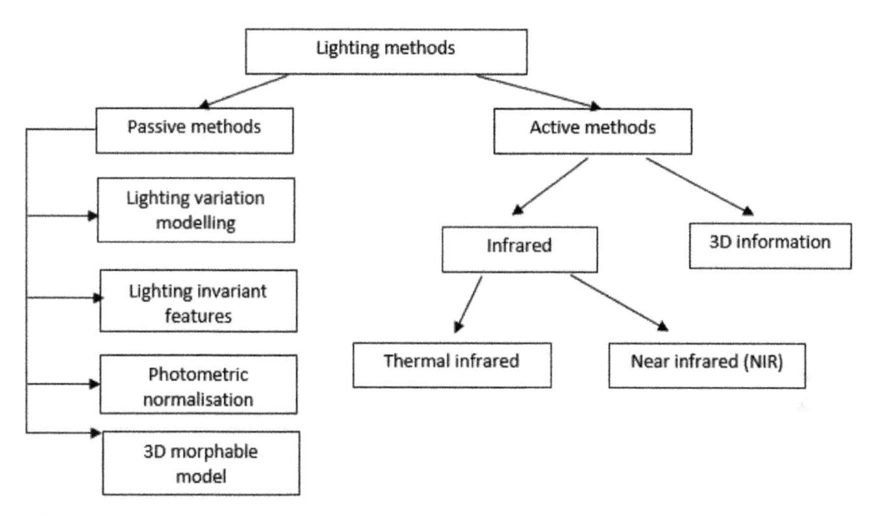

FIGURE 5.1 Categorization of lighting methods.

Passive methods are used to overcome the problem of lighting. Inclusive surveys of lighting methods, mainly passive methods, were reported by Xuan et al.[25] and Ramji.[26] Passive method classifies into four groups: light variation model,[27–32] lighting methods features,[18,24,33–39] photometric normalization[40–43] and 3D morphable model.[44,45] The drawback in this method is that we loss so many useful information of the face.

In active methods, dynamic imaging procedures are utilized to overcome light difference. These techniques are utilized to get facial pictures of lighting invariant modalities or to procure facial pictures taken in steady light conditions. This method is divided on bases of 3D information[45–48] and on bases of infrared.[49,50] Infrared is again divided into two parts: thermal infrared[51–62] and near infrared.[63–72] The main drawbacks of this method are increased costs and creation of complex computation when 3D pictures are utilized. The other problems of this method are when it takes thermal images it changes because of environment temperature, health condition, and due to sweat.[46,73] Some other related papers are also present.[119,152–165]

Recently, the thermal face recognition is used in many places because of robustness and quality of images (Fig. 5.2).[74,75] In this we use IR sensor to sense the thermal rays emitted by the object. We use this in our face recognition in the night time surveillance with little or no light. The fact is that the image is formed due to thermal emissions from the skin and the emission depends on the distribution of blood vessels under the skin.

On the other hand, IR face recognition systems still have limitations with temperature variations and recognition of people wearing eye glasses. In this we will fuse IR images with visible images to enhance the performance of face-recognition systems.

Visible	Infrared			
	Near IR	Short wave IR	Medium wave IR	Long wave IR

0.4nm 0.75nn 1.4nm 3nm 8nm 15nm

FIGURE 5.2 Radiation spectrum ranges.

Face recognition has four major modules: face detection, face alignment, face representation, and face matching. Face detection is the first step in any face recognition framework where the facial region is segmented from its background before further processing. Face alignment aims to detect facial feature points. Accurate detection of these points is crucial to the success of the later stages of the pipeline. Face representation is the most important step in face recognition where the extraction of data on different features of the person takes place. This step is challenging from low-resolution face images.

5.2 LITERATURE REVIEW

5.2.1 EIGENFACES

It is one of most common approaches for face recognition that is also known as Eigen picture, eigenvector, and principal component.[76,77] PCA efficiently shows picture of the face. According to PCA any faces' pictures could be roughly recreated by a little gathering of weights for each face and a standard face picture (Eigen picture). The weights depicting each face are acquired by anticipating the face picture onto the Eigen picture. Chowdhary et al.[78] utilized eigenfaces, which was spurred by the procedure of Kirby and Sirovich, for confront identification and distinguishing proof. Zhao and Yang[79] proposed another technique to register the covariance matrix utilizing three pictures, each was taken in diverse lighting conditions to represent self-assertive enlightenment impacts, if the question is Lambertian. Pentland et al.[80] broadened their initial work on eigenface to eigenfeatures relating to confront segments, for example, eyes, nose, and mouth. They utilized a particular eigenspace that was created from the above eigenfeatures (i.e., eigeneyes, eigennose, and eigenmouth). This strategy would be less touchy to appearance changes than the standard eigenface strategy.

5.2.2 NEURAL NETWORKS

The best thing about neural networks could be nonlinear in network. The feature extraction step in neural network is more efficient than PCA methods. The first artificial neural networks method used single layer adaptive network for face recognition, which is called WISARD, that contains a different system for each put-away individual.[81] The route in developing a neural system structure is complex for creating successful recognition. It is especially a subject to the planned application. For face identification, multilayer perceptron[82] and convolutional neural system[83] have been connected. For face confirmation,[84] a multidetermination pyramid structure. Lawrence et al.[83] proposed a hybrid neural system that consolidates nearby pictures inspecting, a self-sorting out guide (SOM) neural arrange, and a convolutional neural system. The SOM gives a quantization of the picture tests into a topological space where inputs that are close-by in the first space are likewise adjacent in the yield space along these lines giving measurement diminishment and invariance to minor changes in the picture test. The convolutional network extracts progressively bigger highlights in a various leveled set of layers and gives fractional invariance to interpretation, pivot, scale, and twisting. The accuracy of this method is 96.2%.

5.2.3 GRAPH MATCHING

It is another approach for face recognition which is given by Lades and with other colleagues, 1993. Lades et al.[85] introduced a dynamic connection structure for mutilation invariant question acknowledgment that utilized flexible diagram coordinating to locate the nearest stored graph. Dynamic connection engineering is an expansion to the established fake neural networks. Retained items are spoken to by sparse diagrams whose vertices are marked with a multiresolution depiction as far as a neighborhood control range and whose edges are named with geometrical separation vectors. Question acknowledgment can be planned as versatile chart coordinating which is performed by stochastic improvement of a coordinating expense work. They detailed great outcomes on a database of 87 individuals and a little arrangement of office things containing extraordinary rotation of 15.

5.2.4 HIDDEN MARKOV MODELS

This technique is very successful for voice application. This method is also used for human face recognitions.[86] Face is divided into many regions like eyes, nose, lips, etc., which can be linked with hidden Markov models (HMM). HMM need 1D values but images are present in 2D. For that we have to convert it into 1D sequences. In the study by Samaria and Harter,[87] a spatial perception arrangement was extricated from a face picture by utilizing a band sampling method. Each face picture was represented by a 1D vector arrangement of pixel perception. Every perception vector is a square of L lines and there is an M-lines cover between progressive perceptions. An unknown test picture is first examined to a perception grouping. At that point, it is coordinated against each HMMs in the demonstration face database (each HMM speaks to an alternate subject). The match with the most elevated probability is considered the best match and the applicable model uncovers the character of the test face. The accuracy rates were 87% and in HMM 2D was 95%.

5.2.5 GEOMETRICAL FEATURE MATCHING

In this technique calculation of an arrangement of geometrical highlights from the photo of a face. The way that face acknowledgment is conceivable even at coarse determination as low as 8 × 6 pixels[88] when the single facial highlights are scarcely uncovered in detail, infers that the in general geometrical arrangement of the face highlights is adequate for recognition. The general arrangement can be portrayed by a vector speaking to the position and size of the primary facial highlights, for example, eyes and eyebrows, nose, mouth, what's more, the state of face diagram. Geometrical element coordinating in light of accurately estimated separates between highlights might be most valuable for discovering conceivable matches in an expansive database, for example, a Mug shot collection. It will completely depend on the feature location algorithm accuracy. Current computerized confront highlight area calculations do not give a high degree of exactness and require significant computational time. The accuracy rate of this technique is 86–94%.

5.2.6 TEMPLATE MATCHING

In this technique, image is converted into a 2D array that array value is compared with the original image array using Euclidean distance and other formula, single template contains full representation of face. There are so many template matchings that are very good and accurate for face recognition. We can use more than one template for representing different phases of single face. Bruneli and Poggio[89] choose four templates, that is, eyes, nose, mouth, and whole face, and then they compared the geometrical matching and template matching techniques. From that it is clear that template matching was more superior to geometric matching. PCA is a linear combination, the technique cannot achieve better results than correlation.[89] The main drawback of this technique is having complex computation. Template matching is a more logical approach. Presently there is no technique available that is free from limitation.

5.2.7 3D MORPHABLE MODEL

The morphable face show depends on a vector space portrayal of countenances[90] that is developed with the end goal that any raised mix of shape and surface vectors of an arrangement of illustrations portrays a practical human face. Fitting the 3D morphable model to pictures can be utilized as a part of two different ways for acknowledgment crosswise over various review conditions: condition 1. In the wake of fitting the model, acknowledgment can be based on show coefficients, which speak to inherent shape and surface of appearances, and free of the imaging conditions 2. 3D face remaking can be utilized to produce synthetic views from exhibition test pictures.[91–94] This is accomplished by another instatement system that builds power and unwavering quality of the framework, extensively. The new instatement utilizes picture directions of somewhere in the range of six and eight element focuses. The accuracy of this technique is 95.9%.

5.2.8 LINE EDGE MAP

This approach is proposed by Gao and Leung[95] in this face feature are extracted by lines. This approach is a combination of template matching and geometrical feature matching. The line edge map (LEM) approach not just has the benefits of highlight-based methodologies, for example, invariance to brightening and low memory prerequisite, it also has the upside of high acknowledgment execution of template matching. LEM coordinates the basic data with spatial data of a face picture by gathering pixels of face edge guide to line fragments. In the wake of diminishing the edge outline, polygonal line fitting procedure[96] is connected to produce the LEM of a face. The accuracy of this technique is 96.43%. This approach is greater than eigenface for identifying face in different light conditions. It is more sensitive to large facial expression changes.

5.2.9 SUPPORT VECTOR MACHINE

SVM is a learning system that is viewed as a compelling technique that is a broadly useful example for acknowledgment in light of its high speculation execution without the need to include other information.[97] Instinctively, given an arrangement of focuses having a place with two classes, an SVM finds the hyperplane that isolates the biggest conceivable portion of purposes of a similar class on the same side, while boosting the separation from either class to the hyperplane. As indicated by Chowdhary,[97] this hyperplane is called optimal separating hyperplane that limits the danger of misclassifying not just the cases in the preparation set yet in addition the inconspicuous case of the test set. The main properties of SVM are (1) they limit a formally demonstrated upper bound on the speculation blunder; (2) they take a shot at high-dimensional highlight spaces by methods for a double detailing regarding portions; (3) the forecast depends on hyperplanes in these element spaces, which may relate to very included characterization criteria on the information; and (4) that outliers in the preparation informational collection can be taken care of by methods for delicate edges.

5.2.10 HOLISTIC METHODS

In Holistic methods, we take the face image, so it does not support to process the feature separately. This is a unique method in functioning for face image and help in processing of face in a different way in compare to other.[98] There are so many researchers who used this approach for face recognition. In 1992, Prokoski et al.[99] studied the IR imaging for the FR by extracting the significant shape called as "elementary shapes" from the thermograms and the structure of these elementary shapes was like the finger prints. The different methods were used to extract these elementary shapes from the thermograms, but no technical details of these methods are completely available and also no published work is found to show effectiveness of these methods.

5.2.11 CLASSICAL METHODS

Earlier researcher used Holistic method and followed the work of Prokoski et al.[99] The first study for IR images used by Cutler[100] and introduced the method eigen face that was proposed by Pentland.[101] Cutler got the accuracy of face recognition rate approximately 96% by using a database of 288 thermal images, consisting of 12 images for each 24 subject and the images in data base were representing the variation in pose and facial. Later, Socolinsky et al.[102] developed different intensify linear methods, such as Eigen faces, linear discriminant analysis (LDA), local feature analysis (LFA), and independent component analysis (ICA) for the thermal and visible data-based FR. They concluded that the accuracy in thermal spectrum is much higher (approximately 93–98%) than VS, even a range of nuisance variables were present in the data base images.

5.2.12 CONTEMPORARY METHODS

The face recognition methods in IR have similarities with VS but less complex in comparison to it. In 2011, Elguebaly and Bouguila[103] proposed a method based on a Gaussian mixture model, for which the parameters were taken from the sample image using Bayesian approach, but this study achieved the FR accuracy of approximately 95% on the thermal/visible database. Lin et al.[104] developed another method for the FR by considering a database of 50 individuals and per individual 10 images. This study found evidence for the support of FR in the IR spectrum.

Mostly, the methods used complex techniques rather than the inclusion of data-specific knowledge. This is a major drawback in this body of research that the data considered for recognition have not included the types of intrapersonal variations due to different emotional states, alcohol intake, or exercise, or even ambient temperature.

5.2.13 FEATURE-BASED METHODS

Different researchers use feature-based methods to extract the features of IR images for face recognition. Features of an IR image are taken out by using local binary patterns (LBP), wavelet transform, curvelet transform, vascular network, and blood perfusion. The wavelet transform is used to represent 1D and 2D signals, including the face appearance in VS. The curvelet transform increases the functioning of wavelet transform in which the degree of orientational localization directly depends on the scale of the curvelet. In year 1997, Yoshitomi et al.[105] took out the features from thermal images and proposed method based on combining the results of neural networks-based classification and locally averaged appearance. The proposed method was executed at the room temperature ranging from 302 K to 285 K and the recognition rate was calculated 92% when the training and test data were captured at the same room temperature. Meanwhile, the method achieved the recognition rate nearly 60% when the temperature difference was kept 17 K among the training and sample data.

5.2.14 MULTIMODAL METHODS

In IR faces there are several challenges of face recognition methods, such as opaqueness of eye glasses and the dependence of the acquired data on the emotional and physical condition. In contrast to this, the eye glasses, as well as the emotional state, do not produce any limitation in the face recognition. Due to several challenging factors, IR and visible spectrums can be pointed to be supportive to each other. Different methods described for face recognition in visible and IR spectrum use the concept of fusion. The fusion of images can be performed by the two techniques. One is based on data level and other is on decision level. In the data level, the features are constructed by inheriting information from both styles, and then the features are classified while within decision level, the accuracy of matching of two individuals in the IR and VS are calculated. Wilder et al.[106] is the first researchers who use the concept of fusion of the two spectrums. The three methods, transform coded gray scale projections, eigenfaces, and pursuit filters, were used for matching the images. Then the comparison was done to check the performance in isolation, as well as in their fusion, and gray scale projections-based method was ranked as the best (Table 5.1). The proposed fusion method had great effect as the error rates are reduced from 10% down to 1%, approximately.

TABLE 5.1 Literature Comparison.

Name	Method	Performance	Disadvantages
Low-dimensional procedure for characterization of human face	PCA	Recognition rate is low	Only single factor can be varied
Eigen face versus Fisherfaces: recognition using class specific linear projection	Fisher's linear discriminant	Recognition rate higher than PCA	Global feature vectors are generated
Recognize face with PCA and ICA	Independent component analysis	Recognition rate is improved compared to PCA and FLD	Computationally expensive than PCA
Two-dimensional PCA: a new approach for appearance based face representation and recognition	2D PCA	Recognition rate is higher than PCA	Storage requirement is higher than PCA
The importance of the color information in face recognition	Global eigen approach using color image	YUV color space has highest recognition rate	RGB color space does not provide any improvement in recognition rate
A novel hybrid approach based on sub-pattern technique and E2DPCA	Subpattern Extended 2D PCA	Recognition rate higher than PCA, 2DPCA	Variation in lighting, pose are not considered
Face Recognition using a color subspace LDA approach	Color subspace linear discriminate analysis	Recognition rate is higher than 2DPCA and LDA	Variation of performance in color space is not evaluated
Multilinear Image Analysis for Facial Recognition	Multilinear image analysis	Recognition rate higher than PCA	Less and high performance than color subspace LDA
Gabor filter based face recognition technique	2D gabor filter bank	Higher recognition rate than PCA, LDA, 2DPCA, global eigen approach	Low and higher frequency component attenuation
Local binary patterns for multiview facial expression recognition	Local Gabor binary pattern	Better recognition rate than Gabor filter bank	Color information is not included

Another problem is that the performance of the algorithm decreases as time will pass without considering the acquisition conditions. It was observed that performance was decreased due to the change in some tangible factors. In year 2004, Siddiqui et al.[107] take the external temperature on face by using thresholding and image enhancement to normalize the face regions. Chen et al.[108] tell the effect of time passes on performance of IR-based Face recognition that show the error affect as time passes in between sample and test data.

Arandjelovi et al.[109] developed the fusion method and provided the optimum weight of matching scores in an illumination specific manner. Moon et al[110] also conducted a study that focuses on the controlling of the involvement of visible and IR image. During the study, they represent the face images by using the coefficients that were acquired by the wavelet decomposition of an input face image. In the years 2005 and 2009, Kwon et al.[110] and Zahran et al.[111] also used the concept of wavelet-based fusion during their work. Other approaches of Multimodal are geometric invariant moment based given by Abas and Ono,[112] elastic graph matching-based method studied by Hizemtextit et al.,[113] isotherm-based method studied by Tzeng et al., etc.

Jindal and Kumar[114] proposed the method that used PCA and neural network for the recognition of face. In this method, feature is extracted using PCA and the dataset is made, then the dataset is placed in network for training and testing purpose.

5.3 COMMON FACE RECOGNITION SYSTEM

We take input image and extract the features of the image and save in the database. For extracting the features of image, we use preprocessing like normalization. At the time recognition we do verification and validation. If we get the match, we show the match results otherwise show an error message (Fig. 5.3). The dataset we get is used in neural network for better result and accuracy.

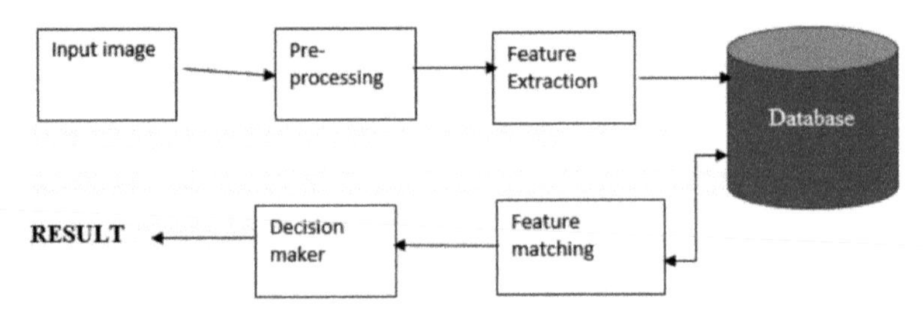

FIGURE 5.3 Architecture diagram of face recognition system.

In face recognition system there are several modules. We will discuss them one by one.

1. Eigen face computation

   This module is used to compute the face space used for face recognition. The recognition is carried out in the face bundle object. But the preparation of such objects requires lots of computation. The steps are as follows:

   - Compute an average face.
   - Build a covariance matrix.
   - Compute eigen values and eigen vector.
   - Select only 16 largest eigen values and its corresponding eigen vectors.
   - Compute the faces using our eigen vectors.
   - Compute eigenspace for our given images.

2. Identification

   - This module does the functionality of taking the image from the above module and then it compares or searches with the images present in database. If any image match is found then a success message is shown to the user.

3. Creating database

   - In this module the input images feature as templates were store in the database and these templates were used for future identification

4. Input image

   - In this module the image of the subject will be taken using camera or from the store folder for verification and validation. It is also used for storing the image of the subject in database.

5. Train the database

   - This module is used to take the stored templates from database to temporary memory or resister to make the system work fast.

6. Preprocessing of image

   - This module improves the image data that suppresses unwilling distortions or enhances some image features important for further processing, although geometric transformations of images.

7. Feature matching
   - In this module we take the user's image features and match with features that are stored in database.
8. Decision or result
   - This module will show the result by completing all the work.

5.4 DIFFERENT TECHNIQUES

Different techniques available for face recognition are as follows:

1. Manual measurement (semiautomated system; 1960): Woodrow Wilson Bledsoe[115] is the father of face recognition. He made a system in which we have to manually put the coordinates of face features, like eyes, nose, mouth, etc., in RAND table. It has rows and columns where we place coordinates then store that table or matrix in database. Whenever the verification of a person is required, the face coordinates are matched with stored matrix.
2. Facial marker (1970): In 1970, Goldstein et al.[116] increase the accuracy of manual face recognition. They take 21 special mark on face like lip thickness, hair color, etc., to increase accuracy. But still it handles manually.
3. Automated System (1973): In 1973, the first automated face recognition system was made by Takeo Kanade.[117] In this, he stored features of face from image in database. Whenever a person comes for verification the face features will be matched with stored features.
4. PCA or eigen face (1987 and 1991): In 1987, Sirovich and Kirby[118] they introduce PCA-based face recognition in which they use linear algebra. This approach used to represent the facial image in low dimension. They successfully show less than 100 values that are used for normalizing the face image.

   In 1991, Turk and Pentland[119,120] saw that they can use eigenface technique for real-time face recognition system. They are the first person who developed this technique and it became very popular in 2001. It helped the arrest of 19 people whose cases were pending. PCA discover the least mean squared blunder direct subspace that maps from the first N dimensional information space into an M-dimensional element space. By doing this, eigenfaces (where regularly $M \ll N$) accomplish dimensionality decreases by utilizing the M eigenvectors of the covariance lattice relating to the biggest eigenvalues. The subsequent premise vectors are acquired by finding the ideal premise vectors that augment the aggregate change of the anticipated data (i.e., the set of premise vectors that best depict the information).

5. Template based (1989): It was proposed by Yuille et al.[121] In this author extracted the facial features using energy function. Every feature was placed in a separate template, likewise a single face has more than one template. By this accuracy of verification was increased.
6. LDA (1996): It was introduced by Swets and Weng[122] for face recognition. It is the improved version of PCA. In this, the face image is treated like a class. It is an appearance-based method. It uses discriminating power at maximum level for feature selection. It is a statistical approach to check unknown classes with known classes. When managing high-dimensional face information, this strategy faces the little example measure issue that emerges where there are few accessible preparing tests looked at to the dimensionality of the example space.
7. Elastic bunch graph matching (EBGM, 1997): EBGM technique first proposed by Wiskott et al.[123] and it became very popular. This approach is based on Gabor wavelets. In this, image of face is represented in graph where all nodes are connected to each other. Each node represents the features of face, like nose, eyes, mouth, etc. An arrangement of preparing pictures is represented by the comparing cluster of picture diagrams of those pictures. An arrangement of complex Gabor wavelet coefficients (or Gabor jets) are utilized as local features at every node. These Gabor jets contain data of various introductions and frequencies for every node. When performing verification of new face image, each graph in the preparation set is coordinated to the picture and the best match demonstrates the personality of an individual.
8. SVM (1997): It was first discovered by Osuna et al.[124] and it became a very successful and important technique in face recognition. SVM is one of the most valuable strategies in characterization issues. One clear illustration is confronted acknowledgment. Be that as it may, SVM cannot be connected when the component vectors characterizing tests have missing sections. An order calculation that has effectively been utilized in this structure is the all-known SVM[124]

that can be connected to the first appearance space or a subspace of it got in the wake of applying an element extraction technique.[124–126] The upside of SVM classifier over customary neural system is that SVMs can accomplish better speculation execution.

9. Fisherface (1997): This technique was discovered by Belhumeur et al.[127] for face recognition. They use discriminate analysis for face recognition. The Fisherface strategy is an improvement of the Eigen face technique that utilizes Fisher's linear discriminant analysis for the dimensionality decrease. The Fisher face expands the proportion of between class disperse to that of inside class scramble, in this way, it works superior to PCA for motivation behind separation. The Fisher face is particularly helpful when facial pictures have huge varieties in enlightenment and outward appearance.
10. Neural network (1998): Rowley et al.[128] were first to use neural network for face recognition. Neural network propelled by human mind made out basic fake neurons likewise known a perceptron's are associated with each other in different layer. Each perceptron comprises of numerical capacity either a summation function (sum up every one of the sources of info and pass) or edge function (limit the information). This is a self-learning network which is prepared and not expressly programmed.[128] On account of face identification, neural network framework analyzes every single window (significantly little in estimate) to decide if it comprises face or not. It diminishes the computational assignment as it does not require to prepare with nonconfront images.[129] This procedure is isolated into two stages. In the initial step, area of the picture (20×20-pixel estimate) is encouraged as contribution to channel made up of the neural network. The yield of this channel lies between [−1, 1] portraying the nonattendance or nearness of face. The channel is connected to every one of the areas of the picture for recognition of countenances. The second step is to conquer the false recognitions found in the initial step and to expand the productivity for better outcomes. This is conceivable if all the covering locations of the single neural network are combined.
11. 3D morphable face model (3DMFM, 1999): It is introduced by Blanz and Vetter[130] for face recognition. 3D morphable models are utilized to confront investigation in light of the fact that the inborn properties of 3D faces gives a portrayal that is insusceptible to intraindividual varieties, for example, posture and brightening.

It is constituted by two PCA-based parametric models, that is, shape and surface models, that are prepared from an arrangement of model 3D confront examines. A 3DMM can create reasonable face cases by controlling its model parameters. What's more, lighting and camera models can be utilized to render such faces with appearance varieties in posture and enlightenment. By using a 3DMM on a 2D picture, we can recover the 3D shape and surface data and gauge the scene properties (light and camera demonstrate parameters). Inferable from these focal points, 3DMM has been broadly utilized in numerous zones including, yet not constrained to, design acknowledgment.[130–133]

12. Active shape model (2000): It is also known as active appearance model that was discovered by Cootes et al.[134] An active appearance model (AAM) is an incorporated measurable model that consolidates a model of shape variety with a model of the appearance varieties in a shape-standardized casing. An AAM contains a measurable model of the shape and dark level appearance of the question of intrigue that can sum up any legitimate illustration. Coordinating to a picture includes discovering model parameters that limit the contrast between the picture and an incorporated model illustration anticipated into the picture.
13. Bayesian method (2000): It first introduced by Moghaddam et al.[135] for face recognition. A probabilistic closeness measure in light of Bayesian conviction that the picture power contrasts are normal for ordinary varieties in appearance of a person. Two classes of facial picture varieties are characterized: intrapersonal varieties and extra personal varieties. Closeness among faces is estimated utilizing Bayesian run the show.
14. Fourier transform (2001): It was first introduced by Norbert wiener for face recognition. The Fourier transform is a critical picture handling device that is utilized to break down a picture into its sine and cosine segments. The output of the picture in the Fourier or frequency space, while the input picture is the spatial area proportional. In the Fourier space picture, each point speaks to a specific frequency contained in the spatial area picture. The Fourier transform is utilized in an extensive variety of utilizations. The Fast Fourier Transform is a productive calculation to figure the discrete Fourier Transform (DFT) and its opposite. The DFT changes one capacity into another, which is known as the frequency

space portrayal of the first capacity. The DFT requires an information work that is discrete. Such sources of info are regularly made by examining a consistent capacity. The DFT is the tested Fourier Transform and hence does not contain all frequencies framing a picture, but rather just an arrangement of tests which is sufficiently expansive to completely portray the spatial space picture. The quantity of frequencies compares to the quantity of pixels in the spatial space picture, that is, the picture in the spatial and Fourier spaces are of a similar size.

15. ICA (2002): It was first introduced by Marian et al.[136] for face recognition. It limits both second-arrange and higher-arrange conditions in the input information and endeavors to discover the premise along which the information (when anticipated onto them) are factually free. Bartlett and with other colleagues given two models of ICA to confront acknowledgment errand: Architecture I—factually free premise pictures, and Architecture II—factorial code portrayal. It is a strategy for finding basic variables or parts from multivariate (multidimensional) measurable information. There have to actualize confront acknowledgment framework utilizing ICA for facial pictures having face introductions and diverse enlightenment conditions, which will give better outcomes as contrasted and existing frameworks.
16. Video indexing (2002): It was first introduced by Acosta et al. for video ordering one edge for each second of the video grouping is utilized for confront location. The picture locales containing a face are separated from the edges. On these pictures the DCT highlights for the face acknowledgment are figured utilizing a square size that gives around a similar measure of highlights vectors for all face pictures. The element exhibits are then assembled by the HMM-bunching. The greatest bunches contain the primary individuals of the video arrangement and the event in the video succession can be additionally assessed.[137]
17. HMM (2003): It is introduced by Perronnin and Dugelay[138] for face recognition. Concealed Markov Models (HMM) are an arrangement of measurable models used to portray the factual properties of a flag. It comprises two interrelated procedures: (1) a hidden, imperceptible Markov chain with a limited number of states, a state change likelihood grid and underlying state likelihood dissemination and

(2) an arrangement of likelihood thickness capacities related with every state. Detection execution of a one-dimensional HMM for dark scale confront pictures. For frontal face pictures, the huge facial locales (hair, temple, eyes, nose, and mouth) arrive in a characteristic request start to finish, regardless of whether the pictures experience little turns in the picture plane and additionally turns in the plane opposite to the picture plane. Every one of these facial locales is allocated to a state in a left to right ID consistent HMM.[139]

18. 3D Face Recognition (2003): It is first introduced by Bronstein and his colleagues[140] for Face recognition. The fundamental curiosity of this approach is the capacity to look at surfaces autonomous of characteristic disfigurements coming about because of outward appearances. To start with, the range picture and the surface of the face are procured. Next, the range picture is preprocessed by expelling certain parts, for example, hair, which can confuse the acknowledgment procedure. At last, a standard type of the facial surface is registered. Such a portrayal is harsh to head introductions and outward appearances, in this manner fundamentally disentangling the acknowledgment system. The acknowledgment itself is performed on the standard surfaces.
19. LBP (2004): It is first proposed by Ahonoen et al.[141] for face recognition. The LBP operator[142] is a ground-breaking method for surface depiction. It is invariant regarding monotonic gray scale changes, consequently no gray scale standardization should be done before applying the LBP operator. The operator names the pixels of a picture by thresholding the area of every pixel with the middle esteem and thinking about the outcome as a paired number. At that point the histogram of the names can be utilized as a surface descriptor. Utilizing roundabout neighborhoods and bilinear introduction at whatever point the examining point does not fall in the focal point of a pixel permits the utilization of any sweep and number of testing focuses. Supposed uniform examples[142] can be utilized to diminish the quantity of containers in the histogram.
20. Scale-invariant feature transform (SIFT, 2004): It was first published by David Lowe in 1992 and first developed by Lowe[143] for face recognition. SIFT has been proposed for separating particular invariant highlights from pictures to perform coordination of various perspectives of a protest or scene. It comprises

two fundamental parts: Intrigue point indicator and highlight descriptor. The SIFT strategy initially identifies the nearby key-focuses that are prominent and stable for pictures in various goals and utilizations scale and pivot invariant descriptors to speak to the key-focuses. In this regard, SIFT highlights are very comparable with LBP highlights with nearby histogram designs speaking to the entire face picture. In spite of the fact that SIFT has great execution in protest acknowledgment, regardless of whether it is a decent descriptor for confront pictures ought to be broke down additional. Since question acknowledgment requires just coarse highlights while confront acknowledgment needs substantially more unpretentious and refined discriminative highlights. An examination of SIFT includes on confront portrayal has ever been done as the primary endeavor to break down the SIFT approach in confront investigation setting.

21. Laplacianfaces (2005): It was first introduced by He et al.[144] for face recognition. Laplacianfaces is an appearance-based technique to deal with human face portrayal and acknowledgment. The approach utilizes Locality Preserving Projection to take in a locality preserving subspace that seems to catch the inborn geometry of the information and the neighborhood structure. At the point when the projection is gotten, each face picture in the picture space is mapped to the low-dimensional face subspace, which is portrayed by an arrangement of highlight pictures, they are called Laplacianfaces. Laplacianfaces technique expects to protect the neighborhood structure of the picture space. It considers the complex structure which is demonstrated by a nearness chart. In some certifiable characterization issues, the nearby complex structure could easily be compared with the worldwide Euclidean structure, particularly when closest neighbors, like classifiers, are utilized for order.
22. Histogram of oriented gradients (HOG, 2005): This technique was first introduced by Dalal et al.[145] for face recognition. The HOG is a component descriptor utilized in PC vision and picture preparing with the end goal of protest location. The strategy includes event of angle introduction-restricted segments of a picture. This strategy is like that of edge introduction histograms, scale-invariant element changes descriptors, and shape settings; however, varies in that it

is registered on a thick matrix of consistently dispersed cells and utilizations covering neighborhood differentiate standardization for enhanced precision.

23. LFA (2006): This technique first used by Ersi et al.[146] for face recognition. LFA separates highlight from picture squares. Before preparing of picture, we partitioned the picture into squares, at that point we utilized all square pictures as the preparing picture. To all square pictures, we built the highlight space by methods for PCA. At last, we did the face acknowledgment as indicated by utilizing the nearby component vectors of all square pictures as acknowledgment include.
24. Hybrid model (2009): It was first used by Wang et al.[147] for face recognition. Under the hybrid approach the merge local properties and entire properties is utilized. Particular eigenface, hybrid nearby component techniques are for hybrid approach. Human facial element plays an essential role in confronting acknowledgment. Research and studies have established that eyes, mouth, what's more, nose are among the most noteworthy elements for acknowledgment.
25. Structural model (2013): This technique was first used by Yan et al.[148] for face recognition. Structural face demonstrates empowers part subtype choice to portray appearance varieties of the nearby part, and part distortion to catch the deformable varieties between various stances and styling. During the time spent identification, the information competitor are first fitted by the structural model to conclude the part area and part subtype, and the certainty score is then processed in view of the fitted design to lessen the impact of structure variety.
26. DeepFace (2014): This technique was first introduced by Taigman et al.[149] for face recognition. After the disappointment of its photograph labeling highlight, Facebook quit fooling around about this innovation and began its exploration in the Facebook's artificial intelligence (AI) lab and named it the DeepFace. This innovation depends on machine learning sub-branch called Deep Learning. It recognizes human faces in computerized pictures. It utilizes a nine-layer neural net with more than 120 million association weights, and was prepared on 4 million pictures transferred by Facebook users. The fascinating thing is that 4 million pictures used to prepare this framework were taken from the profile of only

4030 dynamic Facebook clients! The exactness of this framework is 97.35%, which is as yet 0.28% not as much as that of a human.

27. Convolutional neural network (CNN; 2015): This technique was first introduced by Li et al.[150] for face detection. CNNs is a ground-breaking image processing, AI that used profound figuring out how to perform both generative and distinct assignments, frequently utilizing machine vision that incorporates image and video acknowledgment, alongside recommender frameworks and natural language processing. In CNN each system layer goes about as an identification channel for the nearness of particular highlights or examples exhibit in the first information. The primary layers in a CNN identify (substantial) highlights that can be perceived and deciphered generally simple. Later layers distinguish progressively (littler) highlights that are more dynamic (and are normally present in a significant number of the bigger highlights recognized by before layers). The last layer of the CNN can make an ultraparticular characterization by consolidating all the particular highlights identified by the past layers in the information.
28. FaceNet (2015): This was introduced by Schroff et al.[151] for face recognition. In June 2015, Google ran one better with Face Net, another acknowledgment framework with unrivaled scores: 100% exactness in the reference test Labeled Faces in The Wild, and 95% on the YouTube Faces DB. Utilizing a counterfeit neural system and another calculation, the organization from Mountain View has figured out how to connect a face to its proprietor with relatively idealize results. This innovation is fused into Google Photos and used to sort pictures and consequently label them in light of the general population perceived. Demonstrating its significance in the biometrics scene, it was immediately trailed by the online arrival of an informal open-source adaptation known as OpenFace. Directly takes in a mapping from confront pictures to a minimal Euclidean space where separates straightforwardly compare with a proportion of face likeness. When this space has been delivered, errands, for example, confront acknowledgment, check, and bunching can be effortlessly executed utilizing standard procedures with FaceNet embeddings as highlight vectors. Our technique utilizes a profound convolutional arrange prepared to straightforwardly improve the implanting itself, instead of a transitional bottleneck layer as in

past profound learning approaches. To prepare, we utilize triplets of generally adjusted coordinating/noncoordinating face patches created utilizing a novel online triplet mining technique.

5.5 COMPARISON

TABLE 5.2 Techniques Comparison.

Technique	Authors	Advantage	Disadvantage
PCA (1988–1991)	Turk and Pentland	Decrease the dimension of image in short time, Simple to implement	Learning is time consuming, very sensitive to scale
Template based (1989)	Yuile et al.	100% recognition rate, Simple manner	Complexity between image and template for long time
LDA (1996)	Swets and Weng	98% accuracy rate	Required good quality of image and big database
EBGM (1997)	Wiskott et al.	96% accuracy rate	Very sensitive to lightening condition and need lots of graphs
SVM (1997)	Osuna et al.	Better computational speed and memory, Minimum error rate 8.79%	Worse performance compares to deep learning
Fisherface (1997)	Belhumeur et al.	Better than PCA, 91.5% accuracy rate on Yale face database	More processing time for recognition
Neural Network (1998)	Rowley et al.	Work better on complex problem, accuracy rate 92.33%	Required lots of training
3DMFM (1999)	Blanz and Votter	92.4% accuracy rate on FERET, Recover 3D shape and texture	Difficult construction process
Active Shape Model (2000)	Cootes et al.	Fast, simple, accurate, and efficient to 3D	Treat local model as independent, sparse use of image information
Bayesian Method (2000)	Moghaddan et al.	Interpolation and 90.5% accuracy rate on optical flow	Not automatic, no way to choose prior
Fourier Transform (2001)	Norbert Wiener		
ICA (2002)	Marian et al.	More powerful data representation than PCA	Cannot rank order of dominant component

TABLE 5.2 *(Continued)*

Technique	Authors	Advantage	Disadvantage
Video Indexing (2002)	Acosta et al.	Recognize face from video	Same features vectors (25×30) required for all face
HMM (2003)	Perronnin and Dugelay	Strong statistical foundation, accuracy rate 85–90%	Large number of unstructured parameters, high computational problem
3D face recognition (2003)	Bronstein	Better in capturing surface geometry of face, accuracy rate 93%	Challenging
LBP (2004)	Ahonoen et al.	Maintain high level under localization effect errors	Binary data is sensitive to noise
SIFT (2004)	Lowe	Classical approach, better than other descriptor, accuracy 92.5%	Mathematically complicated, based on HOG
Laplacianface (2005)	He et al.	Preserves local structure and have high discriminative power compare to PCA, 11.4% error rate on Yale database	Less accurate, only pgm file used
HOG (2005)	Dalal et al.	90.2% accuracy rate on	Descriptor vector grows large, take more time to extract and train
LFA (2006)	Ersi et al.	91% accuracy rate on Yale database	Cannot intrinsic distribution
Hybrid Model (2009)	Wang et al.	Combination of two approach	Complex and high computational problem
Structural Model (2013)	Yan et al.	More flexible for representing face	Required high resolution image
Deepface (2014)	Taigman et al.	97% accuracy rate on LFW database	CPU are not optimized
CNN (2015)	Li et al.	Able to learn, weight sharing, good feature extractor, accuracy rate 98.4%	High computational cost, required good GPU, need lots of training
FaceNet (2015)	Schroff et al.	99.63% accuracy rate on LFW database	Required 30–40 images of per person with good quality

5.6 CONCLUSIONS

In this chapter, we review significant number of studies to give a brief description about different face recognition techniques and the recent developments in this field. From the present study there are so many algorithms which are proposed and implemented to enhance the face recognition system. The list of references gives more detailed knowledge about different approaches and techniques.

KEYWORDS

- **face recognition**
- **eigenface**
- **neural network**
- **hidden Markov model**
- **support vector machine**

REFERENCES

1. Bruner, J. S.; Tagiuri, R. *The Perception of People;* 1st ed; Addison: Wesley, USA, 1954.
2. Bledsoe, W. W. *The Model Method in Facial Recognition;* Panoramic Research Inc.: Palo Alto, CA, 1966.
3. Turk, M.; Pentland, A. *Face Recognition Using Eigenfaces*, IEEE Computer Society Conference on Computer Vision and Pattern Recognition, IEEE, Maui, HI, USA, 1991, pp 586–591.
4. Feng, G.; Yuen, P.; Dai, D. Human Face Recognition Using PCA on Wavelet Sub-band. *J. Electron. Imaging* **2000,** *9*, 226–233.
5. Garcia, C.; Zikos, G.; Tziritas, G. Wavelet Packet Analysis for Face Recognition Image. *Vis. Comput.* **2000,** *18*, 289–297.
6. Moghaddam, B.; Jebara, T.; Pentland, A. Bayesian Face Recognition. *Pattern Recognit.* **2000,** *33*, 1771–1782.
7. Bartlett, M. S.; Movellan, J. R.; Sejnowski, T. J. Face Recognition by Independent Component Analysis. *IEEE Trans. Neural Netw.* **2002,** *13*, 1450–1464.
8. Chengjun, L.; Wechsler, H. Gabor Feature Based Classification Using the Enhanced Fisher Linear Discriminant Model for Face Recognition. *IEEE Trans. Image Process.* **2002,** *11*, 467–476.

9. Chien, J-T.; Chia-Chen, W. Discriminant Wavelet Faces and Nearest Feature Classifiers for Face Recognition. *IEEE Trans. Pattern Anal. Mach. Intell.* **2002,** *24,* 1644–1649.
10. Haddadnia, J.; Ahmadi, M.; Faez, K. An Efficient Feature Extraction Method with Pseudo-Zernike Moment in RBF Neural Network-based Human Face Recognition System. EURASIP *J. Adv. Signal Process.* **2003,** *1,* 890–901.
11. Haddadnia, J.; Faez, K.; Ahmadi, M. An Efficient Human Face Recognition System Using Pseudo Zernike Moment Invariant and Radial Basis Function Neural Network. *Int. J. Pattern Recognit. Artif. Intell.* **2003,** *17,* 41–62.
12. Chengjun, L. Gabor-based Kernel PCA with Fractional Power Polynomial Models for Face Recognition. *IEEE Trans. Pattern Anal. Mach. Intell.* **2004,** *26,* 572–581.
13. Chen, W.; Er, M. J., Wu, S. PCA and LDA in DCT Domain. *Pattern Recognit. Lett.* **2005,** *26,* 2474–2482.
14. Ahonen, T.; Hadid, A.; Pietikainen, M.; Member, S. Face Description with Local Binary Patterns: Application to Face Recognition. *IEEE Trans. Pattern Anal. Mach. Intell.* **2006,** *28,* 2037–2041.
15. Liu, C. -C. L.; Dai, D. -Q. D.; Yan, H. Y. Local Discriminant Wavelet Packet Coordinates for Face Recognition. *J. Mach. Learn. Res.* **2007,** *8,* 1165–1195
16. Jadhav, D. V.; Holambe R. S. Radon and Discrete Cosine Transforms Based Feature Extraction and Dimensionality Reduction Approach for Face Recognition. *Signal Process.* **2008,** *88,* 2604–2609.
17. Han, P. Y.; Jin, A. T. B.; Abas, F. S. Neighbourhood Preserving Discriminant Embedding in Face Recognition. *JVCIR* **2009,** *20,* 532–542.
18. Jadhav, D. V.; Holambe, R. S. Rotation, Illumination Invariant Polynomial Kernel Fisher Discriminant Analysis Using Radon and Discrete Cosine Transforms Based Features for Face Recognition. *Pattern Recog. Lett.* **2010,** *31,* 1002–1009.
19. Singh, C.; Mittal, N.; Walia, E. Face Recognition Using Zernike and Complex Zernike Moment Features. *Pattern Recognit. Image Anal.* **2011,** *21,* 71–81.
20. Li, D.; Tang, X.; Pedrycz, W. Face Recognition Using Decimated Redundant Discrete Wavelet Transforms. *Mach. Vis. Appl.* **2012,** *23* (2012), 391–401
21. Singh, C.; Sahan, A. M., Face Recognition Using Complex Wavelet Moments. *Opt. Laser Technol.* **2013,** *47,* 256–267
22. Yan, Y.; Wang, H.; Suter, D. Multi-subregion Based Correlation Filter Bank for Robust Face Recognition. *Pattern Recognit.* **2014,** *47,* 3487–3501.
23. Socolinsky, D. A.; Selinger, A.; Neuheisel, J. D. Face Recognition with Visible and Thermal Infrared Imagery. *Comput. Vis. Image Underst.* **2003,** *91,* 72–114.
24. Chen, H. F.; Belhumeur, P. N.; Jacobs, D. W. *In Search of Illumination Invariants,* IEEE Conference on Computer Vision and Pattern Recognition, IEEE, USA, 2000, pp 254–261.
25. Xuan, Z.; Josef, K.; Kieron, M. *Illumination Invariant Face Recognition: A Survey,* First IEEE International Conference on Biometrics: Theory, Applications, and Systems, IEEE, Washington, DC, USA, 2007, pp 1–8.
26. Ramji, M. Illumination Invariant Face Recognition: A Survey of Passive Methods. *Procedia Comput. Sci.* **2010,** *2,* 101–110

27. Zhang, L.; Samaras, D. *Face Recognition Under Variable Lighting Using Harmonic Image Exemplars*, IEEE Computer Society Conference on Computer Vision and Pattern Recognition, IEEE, Madison, WI, USA, 2003, pp 19–25.
28. Hallinan, P. W. *A Low-dimensional Representation of Human Faces for Arbitrary Lighting Conditions*, IEEE Computer Society Conference on Computer Vision and Pattern Recognition, IEEE, Seattle, WA, USA, 1994, pp 995–999.
29. Shashua; On Photometric Issues in 3D Visual Recognition from a Single 2D Image. *Int. J. Comput. Vis.* **1997,** *21* (1997), 99–122.
30. Georghiades, A. S.; Belhumeur, P. N.; Kriegman, D. From Few to Many: Illumination Cone Models for Face Recognition Under Variable Lighting and Pose. *IEEE Trans. Pattern Anal. Mach. Intell.* **2001,** *23*, 643–660.
31. Zhang, L.; Samaras, D. Face Recognition from a Single Training Image Under Arbitrary Unknown Lighting Using Spherical Harmonics. *IEEE Trans. Pattern Anal. Mach. Intell.* **2006,** *28*, 351–363.
32. Zhou, S. K.; Aggarwal, G.; Chellappa, R.; Jacobs, D. W. Appearance Characterization of Linear Lambertian Objects, Generalized Photometric Stereo, and Illumination-Invariant Face Recognition. *IEEE Trans. Pattern Anal. Mach. Intell.* **2007,** *29*, 230–245.
33. Yongsheng, G.; Leung, M. K. H. Face Recognition Using Line Edge Map. *IEEE Trans. Pattern Anal. Mach. Intell.* **2002,** *24*, 764–779.
34. Wei, S. -D.; Lai, S. -H. In *Robust Face Recognition Under Lighting Variations,* Proceedings of the 17th International Conference on Pattern Recognition, IEEE, Cambridge, England, 2004, pp 354–357.
35. Thomas, Y. C. -H.; Shang-Hong, L.; Long-Wen, V. *Robust Face Image Matching Under Illumination Variations* EURASIP *J. Adv. Signal Process* **2004,** 2533–2543.
36. Shashua, Riklin-Raviv, T., *The Quotient Image: Class-Based Re-rendering and Recognition with Varying Illuminations. IEEE Trans. Pattern Anal. Mach. Intell.* **2001,** *23*, 129–139.
37. Chen, C. -P.; Chen, C. -S. *Lighting Normalization with generic Intrinsic Illumination Subspace for Face Recognition,* 10th IEEE International Conference on Computer Vision, IEEE, Beijing, China, 2005, pp 1089–1096.
38. Zhang, Y.; Tian, J.; He, X.; Yang, X. *MQI Based Face Recognition Under Uneven Illumination,* International Conference on Biometrics, Springer, Seoul, Korea, 2007, pp 290–298.
39. Jadhav, D., Holambe, R. S. *Feature Extraction and Dimensionality Reduction Using Radon and Fourier Transforms with Application to Face Recognition,* International Conference on Computational Intelligence and Multimedia Applications, IEEE, Sivakasi, Tamil Nadu, India, 2007, pp 254–260.
40. Shan, S.; Gao, W.; Cao, B.; Zhao, D. *Illumination Normalization for Robust Face Recognition Against Varying Lighting Conditions*, IEEE International Workshop on Analysis and Modeling of Faces and Gestures, IEEE, Nice, France, 2003, pp 157–164.
41. Du, S.; Ward, R. *Wavelet-based Illumination Normalization for Face Recognition,* IEEE International Conference on Image Processing, ICIP, IEEE, Genoa, Italy, 2005, 954–957.
42. Xie, X.; Lam, K. -M. *An Efficient Illumination Normalization Method for Face Recognition. Pattern Recognit. Lett.* **2006,** 27, 609–617.

43. Chowdhary, C. L.; Acharjya, D. P., In *Breast Cancer Detection using Intuitionistic Fuzzy Histogram Hyperbolization and Possibilitic Fuzzy c-mean Clustering algorithms with texture feature based Classification on Mammography Images*, Proceedings of the International Conference on Advances in Information Communication Technology & Computing, 2016, 21.
44. Blanz, V.; Vetter, T. Face Recognition Based on Fitting a 3D Morphable Model. *IEEE Trans. Pattern Anal. Mach. Intell.* **2003,** *25* (2003), 1063–1074.
45. Bowyer, K. W.; Chang, K. I.; Flynn, P. J., *A Survey of Approaches to Three-dimensional Face Recognition*, International Conference on Pattern Recognition, IEEE, Cambridge, England, 2004, pp 358–361.
46. Bowyer, K. W.; Chang, K.; Flynn, P. *A Survey of Approaches and Challenges in 3D and Multi-modal 3D 2D Face Recognition. Comput. Vis. Image Underst.* **2006,** *101*, 1–15.
47. Scheenstra, A.; Ruifrok, R. C.; Veltkamp. *A Survey of 3D Face Recognition Methods Audio-and Video-Based Biometric Person Authentication;* Springer: Hilton Rye Town, NY, USA, 2005, pp 891–899.
48. Kittler, J.; Hilton, A.; Hamouz, M.; Illingworth, J. *3D Assisted Face Recognition: A Survey of 3D Imaging, Modelling and Recognition Approaches,* IEEE Computer Society Conference on Computer Vision and Pattern Recognition, IEEE, San Diego, CA, USA, 2005, p 114.
49. Ghiass, R. S.; Arandjelović, O.; Bendada, A.; Maldague, X. In *Infrared Face Recognition: a Literature Review,* Proceedings of the International Joint Conference on Neural Networks, IEEE, Dallas, Texas, USA, 2013, pp 1–10.
50. Ghiass, R. S.; Arandjelović, O.; Bendada, A.; Maldague, X. Infrared Face Recognition: A Comprehensive Review of Methodologies and Databases. *Pattern Recognit.* **2014,** *47*, 2807–2824.
51. Socolinsky, D. A.; Selinger, A. In *Thermal Face Recognition in An Operational Scenario*, Proceedings of the 2004 IEEE Computer Society Conference on Computer Vision and Pattern Recognition, IEEE, Washington DC, USA, 2004, pp 1012–1019.
52. Chen, X.; Flynn, P. J.; Bowyer, K. W. In *Visible-light and Infrared Face Recognition,* The Proceedings of Workshop on Multimodal User Authentication, IEEE, Santa Barbara, CA, USA, 2003, pp 48–55.
53. Kong, S. G.; Heo, J.; Abidi, B. R.; Paik, J.; Abidi, M. A. Recent Advances in Visual and Infrared Face Recognition-A Review. *Comput. Vis. Image Underst.* **2005,** *97*, 103–135.
54. Buddharaju, P.; Pavlidis, I. T.; Kakadiaris, I. *Face Recognition in the Thermal Infrared Spectrum*, Conference on Computer Vision and Pattern Recognition Workshop, 2004, pp 133–138.
55. Buddharaju, P.; Pavlidis, I. T.; Tsiamyrtzis, P. *Physiology-based Face Recognition*, IEEE Conference on Advanced Video and Signal Based Surveillance, IEEE, Como, Australia, 2005, pp 354–359.
56. Buddharaju, P.; Pavlidis, I. T.; Tsiamyrtzis, P.*Pose-invariant Physiological Face Recognition in the Thermal Infrared Spectrum*, Conference on Computer Vision and Pattern Recognition Workshop, 2006, pp 53–60.

57. Buddharaju, P.; Pavlidis, I. T.; Tsiamyrtzis, P.; Bazakos, M. *Physiology-based Face Recognition in the Thermal Infrared Spectrum, IEEE Trans. Pattern Anal. Mach. Intell.* **2007,** *29*, 613–626.
58. Abas, K. H. Ono, O. *Implementation of Multi-centroid Moment Invariants in Thermal-based Face Identification System, Amer. J. Appl. Sci.* **2010,** *7*, 283–289.
59. Farokhi, S.; Shamsuddin, S. M.; Flusser, J.; Sheikh, U. U. *Assessment of Time-lapse in Visible and Thermal Face Recognition. World Acad. Sci. Eng. Technol.* **2012,** *62*, 540–545.
60. Seal, A.; Bhattacharjee, D.; Nasipuri, M.; Basu, D. K. *Minutiae Based Thermal Face Recognition Using Blood Perfusion Data*, International Conference on Image Information Processing, Himachal Pradesh, India, 2011, pp 1–4.
61. Ghiass, R. S.; Arandjelovic, O.; Bendada, H.; Maldague, X. In *Vesselness Features and the Inverse Compositional AAM for Robust Face Recognition Using Thermal IR,* Proceedings of AAAI Conference on Artificial Intelligence, AAAI, Bellevue, Washington, USA, 2013, pp 357–364.
62. Ghiass, R. S.; Arandjelovic, O.; Bendada, H.; Maldague, X. *Illumination-invariant Face Recognition from a Single Image Across Extreme Pose Using a Dual Dimension, AAM ensemble in the thermal infrared spectrum,* The 2013 International Joint Conference on Neural Networks, 2013, pp 1–10.
63. Pan, P. Z.; Glenn, H.; Manish, P.; Bruce, T. Face Recognition in Hyperspectral Images. *IEEE Trans. Pattern Anal. Mach. Intell.* **2003,** *25*, 1552–1560.
64. Li, S. Z.; Chu, R.; Liao, S.; Zhang, L. Illumination Invariant Face Recognition Using Near-Infrared Images. *IEEE Trans. Pattern Anal. Mach. Intell.* **2007,** 29, 627–639.
65. Zhao, S.; Grigat, R.-R. In *An Automatic Face Recognition System in the Near Infrared Spectrum*, Proceedings of the 4th International Conference on Machine Learning and Data Mining in Pattern Recognition, Springer, Leipzig, Germany, 2005, pp 437–444.
66. Farokhi, S.; Shamsuddin, S. M.; Flusser, J.; Sheikh, U. U.; Khansari, M.; Kourosh, J.-K. Rotation and Noise Invariant Near-Infrared Face Recognition by Means of Zernike Moments and Spectral Regression Discriminant Analysis. *J. Electron. Imaging* **2013,** *22*, 013030-1-013030-11.
67. Farokhi, S.; Shamsuddin, S. M.; Flusser, J.; Sheikh, U. U.; Khansari, M.; Kourosh, J.-K. Near Infrared Face Recognition by Combining Zernike Moments and Undecimated Discrete Wavelet Transform. *Digit. Signal Process.* **2014,** *31*, 13–27.
68. Hizem, W.; Krichen, E.; Ni, Y.; Dorizzi, B.; Garcia-Salicetti, S. *Specific Sensors for Face Recognition*, International Conference on Biometrics, Springer, Hong Kong, China, 2005, pp 47–54.
69. Yang, N.; Krichen, E.; Hizem, W.; Garcia-Salicetti, S.; Dorizzi, B. Active Differential CMOS Imaging Device for Human Face Recognition. *IEEE Signal Process. Lett.* **2006,** *13*, 220–223.
70. Ni, Y.; Yan, X. -L. In *CMOS Active Differential Imaging Device with Single In-pixel Analog Memory,* Proceedings of the 28th European Solid-State Circuits Conference, IEEE, Florence, Italy, **2002,** pp 359–362.
71. Xuan, Z.; Josef, K.; Kieron, M. *Face Recognition Using Active Near-IR Illumination*, British Machine Vision Conference, Citeseer, Oxford, UK, 2005, pp 1–11.

72. Zou, X.; Kittler, J.; Messer, K. *Ambient Illumination Variation Removal by Active Near-IR Imaging*, International Conference on Biometrics, Springer, Hong Kong, China, 2006, pp 19–25.
73. Bebis, G.; Gyaourova, A.; Singh, S.; Pavlidis, I. *Face Recognition by Fusing Thermal Infrared and Visible Imagery. Image Vis. Comput.* **2006,** *24*, 727–742.
74. Li, S. Z.; Chu, R.; Ao, M.,; Zhang, L.; He, R. *Highly Accurate and Fast Face Recognition Using Near Infrared Images;* Zhang, D., Jain A., (Eds.; International Conference on Biometrics, Springer, Hong Kong, China, 2005, pp 151–158.
75. Di, H.; Yi-Ding, W.; Yi-Ding, W. *A Robust Infrared Face Recognition Method Based on AdaBoost Gabor Features*, International Conference on Wavelet Analysis and Pattern Recognition, IEEE, Beijing, China, 2007, pp 1114–1118.
76. Sirovich, L.; Kirby, M. Low-Dimensional Procedure for the Characterisation of Human Faces. *J. Optical Soc. of Am.* **1987,** *4*, 519–524.
77. Kirby, M.; Sirovich, L. Application of the Karhunen- Loève Procedure for the Characterisation of Human Faces. *IEEE Trans. Pattern Anal. Mach. Intell.* **1990,** *12*, 831–835.
78. Chowdhary, C. L.; Sai, G. V. K.; Acharjya, D. P. Decreasing False Assumption for Improved Breast Cancer Detection. *J. Sci. Arts* **2016,** *35* (2), 157–176.
79. Zhao, L.; Yang, Y. H. Theoretical Analysis of Illumination in PCA-based Vision Systems. *Pattern Recog.* **1999,** *32*, 547–564.
80. Pentland, A.; Moghaddam, B.; Starner, T. In *View-Based and Modular Eigenspaces for Face Recognition*, Proceedings of the IEEE CS Conference on Computer Vision and Pattern Recognition, 1994, pp 84–91.
81. Stonham, T. J. Practical Face Recognition and Verification with WISARD. *Aspects of Face Processing* 1984, 426–441.
82. Sung, K. K.; Poggio, T. Learning Human Face Detection in Cluttered Scenes. *Comput. Anal. Image Patterns* 1995, 432–439.
83. Lawrence, S.; Giles, C. L.; Tsoi, A. C.; Back, A. D. Face Recognition: A Convolutional Neural-Network Approach. *IEEE Trans. Neural Networks* **1997,** *8*, 98–113.
84. Weng, J.; Huang, J. S.; Ahuja, N. In *Learning Recognition and Segmentation of 3D Objects from 2D Images*, Proceedings of the IEEE International Conference on Computer Vision, 1993, pp 121–128.
85. Lades, M.; Vorbruggen, J. C.; Buhmann, J.; Lange, J.; Malsburg, C. V. D.; Wurtz, R. P.; Konen, M. Distortion Invariant Object Recognition in the Dynamic Link Architecture. IEEE Trans. Comput. **1993,** *42*, 300–311.
86. Samaria, F.; Fallside, F. Face Identification and Feature Extraction Using Hidden Markov Models. *Image Processing: Theory and Application;* Vernazza, G., Ed.; Elsevier: Chicago, IL, USA, 1998.
87. Samaria, F.; Harter, A. C. *Parameterisation of a Stochastic Model for Human Face Identification*, Proceedings of the Second IEEE Workshop Applications of Computer Vision. 1994.
88. Tamura, S.; Kawa, H.; H. Mitsumoto Male/Female Identification from 8_6 Very Low-resolution Face Images by Neural Network. *Pattern Recog.* **1996,** *29*, 331–335.
89. Bruneli, R.; Poggio, T. Face Recognition: Features Versus Templates. *IEEE Trans. Pattern Anal. Mach. Intell.* **1993,** *15*, 1042–1052.

90. Vetter, T.; Poggio, T. *Linear Object Classes and Image Synthesis from a Single Example Image, IEEE Trans. Pattern Anal. Mach. Intell.* **1997,** *19* (7), 733–742.
91. Beymer, D. Poggio, T. *Face Recognition from One Model View*, Proceedings of the Fifth International Conference on Computer Vision, 1995.
92. Vetter, T.; Blanz, V. In *Estimating Coloured 3D Face Models from Fingle Images: An Example-Based Approach*, Proceedings of the Conference on Computer Vision (ECCV '98), II, 1998.
93. Georghiades, A. S.; Belhumeur, P. N.; Kriegman, D. J. *From few to many: Illumination cone models for face recognition under variable lighting and pose*, IEEE Trans. Pattern Analysis and Machine Intelligence, 23(6), 643-660.
94. Zhao, W.; Chellappa, R. In *SFS Based View Synthesis for Robust Face Recognition* Proceedings of the International Conference on Automatic Face and Gesture Recognition, 2000, pp 285–292.
95. Chowdhary, C. L.; Ranjan, A.; Jat, D. S.; Categorical Database Information-theoretic Approach of Outlier Detection Model. *Annal. Comput. Sci. Series*, 14th Tome 2nd Fasc. – **2016,** 29–36.
96. Leung, M. K. H.; Yang, Y. H. Dynamic Two-strip Algorithm in Curve Fitting. *Pattern Recog.* **1990,** *23*, 69–79.
97. Chowdhary, C. L. (Linear Feature Extraction Techniques for Object Recognition: Study of PCA and ICA. *J. Serbian Soc. Comput. Mech.* **2011,** *5* (1), 19–26.
98. Axelrod, V. The Fusiform Face Area: In Quest of Holistic Face Processing. *J. Neurosci.* **2010,** 8699–8701.
99. Prokoski, F. J., Riedel, R. B., Coffin, J. S. *Identification of Individuals by Means of Facial Thermography*, International Carnahan Conference on Security and Technology, 1992, pp 120–125.
100. Cutler, R. Face Recognition Using Infrared Images and Eigenfaces, Technical Report, University of Maryland, 1996.
101. Turk, M. Pentland, A. Eigen Faces for Recognition. *J. Cognitive Neurosci.* **1991,** *3*, 71–86.
102. Socolinsky, D. A.; Selinger, A. Thermal Face Recognition Over Time. *Pattern Recog.* **2004,** *4*, 187–190.
103. Elguebaly, T.; Bouguila, N. *A Bayesian Method for Infrared Face Recognition*, Machine Vision beyond Visible Spectrum, 2011.
104. Lin, Z. *Infrared face recognition based on compressive sensing and PCA*, IEEE Conference on CSAE, 2011, 2, pp 51–54.
105. Yoshitomi, Y. *Face Identification Using Thermal Image Processing*, Workshop on Robot & Human Communication, 1997, 374–379.
106. Wilder, J.; Phillips, P.; Jiang, C.; Wiener, S. In *Comparison of Visible and Infrared Imagery for Face Recognition*, Proceedings in Second International Conference on Automatic Face and Gesture Recognition, 1996, pp 182–187.
107. Siddiqui, R. In *Face Identification Based on Biological Trait Using Infrared Images After Cold Effect Enhancement and Sunglasses Filtering*, Proceedings of International Conference in Central European Computer Graphics, Visualization and Computer Vision, 2004.
108. Chen, X.; Bowyer, K. In *Visible Light and Infrared FR*, Proceedings in Workshop on Multimodal User Authentication, 2003, 48–55.

109. Arandjelovi, O.; Cipolla, R. *On Person Authentication by Fusing Visual and Thermal Face Biometrics*, IEEE Conference on VSBS, 2006.
110. Arandjelovi, O.; Hammoud, R. I.; Cipolla, R. *Multi-sensory Face Biometric Fusion for Personal Identification*. CVPRW, 2006.
111. Kwon, O. K.; Kong, S. G. *Multiscale Fusion of Visual and Thermal Images for Robust Face Recognition*, International Conference on Computer Intelligence for Homeland Security and Personal Safety, 2005.
112. Abasand, K. H. Ono, O. In *Thermal Physiological Moment Invariants for Face Identification*. Proceedings of International Conference on Signal-Image Technology & Internet based Systems, 2010, pp 1–6.
113. Hizem, W. Face Recognition from Synchronised Visible and Near-infrared Images. *IET Signal Process.* **2009,** 282–288.
114. Jindal, N.; Kumar, V. Enhanced Face Recognition Algorithm using PCA with Artificial Neural Networks. *Int. J. Adv. Res. Comput. Sci. Software Eng.* **2013,** *3* (6), 864–872.
115. Bledsoe, W. W. *The Model Method in Facial Recognition;* Panoramic Research Inc., Palo Alto: CA, Rep. PRI, 1996.
116. Goldstein, A. J.; Harmon, L. D.; Lesk, A. B. Identification of Human Faces. *Proceedings IEEE* **1971,** *59*, 748.
117. Kanade, T. Picture Processing System by Computer Complex and Recognition of Human Faces. Dept. of Information Science, Kyoto University, 1973.
118. Sirovich, L.; Kirby, M. Low-Dimensional Procedure for the Characterization of Human Face. *J. Optical Soc Am.* **1987,** *4* (3), 519–524.
119. Chowdhary, C. L. Appearance-based 3-D Object Recognition and Pose Estimation: Using PCA, ICA and SVD-PCA Techniques. *LAP Lambert Acad.* 1st ed. Germany, 2011, p 76
120. Chowdhary, C. L.; Sai, G. V. K.; Acharjya, D. P. Decrease in False Assumption for Detection Using Digital Mammography. *Comput. Intell. Data Mining* **2015,** *2*, 325–333.
121. Yuille, A. L.; Cohen, D. S.; Hallinan, P. W. In *Feature Extraction from Face Using Deformable Templates*, Proceedings of CVPR, 1989.
122. Swets, D. L.; Weng, J. J. Using Discriminant Eigenfeatures for Image Retrieval. *IEEE Trans. PAMI* **1996,** *18* (8), 831–836.
123. Wiskott, L.; Fellous, J. M.; Kruger, N.; Malsburg, C. V. D. Face Recognition by Elastic Bunch Graph Matching. *IEEE Trans. Pattern Anal. Mach. Intell.* **1997,** *19* (7), 775–779.
124. Osuna, E.; Freund, R.; Giorsit, F. *Training Support Vector Machines: An Application to Face Detection,* Proceedings. of CVPR, 1997, 130–136.
125. Helsele, B.; Serre, T.; Poggio, T. *A Component-based Framework for Face Detection and Identification*, IJCV **2007,** *74* (2), 167–181.
126. Tao, Q.; Chu, D.; Wang, J. Recursive Support Vector Machines for Dimensionality Reduction. *IEEE Trans. NN* **2008,** *19* (1), 189–193.
127. Belhumeur, P. N.; Hespanha, J. P.; Kriegman, D. J. Eigenfaces vs. Fisherfaces: Recognition Using Class Specific Linear Projection. *IEEE Trans. Pattern Anal. Mach. Intel.* **1997,** *19* (7), 711–720.
128. Rowley, H. A.; Baluja, S.; Kanade, T. Neural Network-based Face Detection. *IEEE Trans. Pattern Anal. Mach. Intell.* **1998,** *20* (1), 23–38.

129. Sung, K. K.; Poggio, T. Example-Based Learning for View-Based Human Face Detection. *IEEE Trans. Pattern Anal. Mach. Intell.* **1998,** *20* (1), 39–51.
130. Blanz, V.; Vetter, T. *Face Recognition Based on Fitting a 3D Morphable Model. IEEE Trans. Pattern Anal. Mach. Intell.* **2003,** *25* (9), 1063–1074.
131. Feng, Z. -H.; Hu, G., Kittler, J.; Christmas, W.; Wu, X. -J. Cascaded Collaborative Regression for Robust Facial Landmark Detection Trained Using a Mixture of Synthetic and Real Images with Dynamic Weighting. *IEEE Trans. Image Process.* **2015,** *24* (11) 3425–3440.
132. Hassner, T.; Harel, S.; Paz, E.; Enbar, R. *Effective Face Frontalization in Unconstrained Images, IEEE Computer Society Conference on Computer Vision and Pattern Recognition,* 2015, *7*, pp 4295–4304.
133. Hu, G.; Yan, F.; Chan, C. -H; Deng, W; Christmas, W.; Kittler, J.; Robertson, N. M. *Face Recognition Using a Unified 3d Morphable Model;* ECCV, Springer: Cham, Switzerland, 2016; pp 73–89.
134. Cootes, T. F.; Walker, K.; Taylor, C. J. In *View-Based Active Appearance Models*, Proceedings of the IEEE International Conference on Automatic Face and Gesture Recognition, 2000, pp 227–232.
135. Chowdhary, C. L.; Muatjitjeja, K.; Jat, D.S.*Three-dimensional Object Recognition Based Intelligence System for Identification*, Emerging Trends in Networks and Computer Communications (ETNCC), 2015.
136. Bartlett, M. S., Movellan, J. R. Sejnowski, T. J. Face Recognition By Independent Component Analysis. *IEEE Trans. Neural Networks* **2002,** *13* (6), 1450–1464.
137. Dimitrova, N.; Zhang, H.; Shahraray, B.; Sezan, I.; Huang, T.; Zakhor, A. Applications of Video Content Analysis and Retrieval. *ACM Comput. Surveys* **2002,** *9* (3), 42–55.
138. Perronnin, F.; Dugelay, J. L.; Rose, K.; *Iterative Decoding of Two-dimensional Hidden Markov Models*, In *Acoustics, Speech, and Signal Processing*, Proceedings (ICASSP'03) 2003 IEEE International Conference, 3, III-329, 2003.
139. Samaria, F.; Young, S. HMM Based Architecture for Face Identification. *Image Comput. Vision* **1994,** *12*, 537–583.
140. Bronstein, A. M.; Bronstein, M. M.; Kimmel, R. *Expression-invariant 3D Face Recognition*, Proceedings of the Audio & Video-based Biometric Person Authentication (AVBPA), Lecture Notes in Computer Science 2688, Springer, 2003, 62–69.
141. Ahonen, T.; Hadid, A.; Pietikäinen, M. In *Face Recognition with Local Binary Patterns*, The 8th European Conference on Computer Vision, Springer, 2004.
142. Ojala, T.; Pietikäinen, M.; Mäenpää, T. Multiresolution Gray-scale and Rotation Invariant Texture Classification with Local Binary Patterns. *IEEE Trans. Pattern Anal. Mach. Intell.* **2002,** *24* (7), 971–987.
143. Lowe, D. Distinct Image Features from Scale-invariant Key Points. *Int. J. Comput. Vision* **2004,** *60* (2), 91–110.
144. He, X.; Yan, S.; Hu, Y.; Niyogi, P.; Zhang, H. J. Face Recognition Using Laplacianfaces. *IEEE Trans. Pattern Anal. Mach. Intell.* **2005,** *27* (3), 328–340.
145. Dalal, N.; Triggs, B. *Histograms of Oriented Gradients for Human Detection*, CVPR, 2005.
146. Ersi, E. F.; Zelek, J. S. Local Feature Matching for Face Recognition. In *Computer and Robot Vision;* The 3rd Canadian Conference, 2006, 4.

147. Wang, Y.; Anderson, P. G.; Gaborski, R. S. In *Face Recognition Using a Hybrid Model,* Applied Imagery Pattern Recognition Workshop (AIPRW), 2009, 1–8.
148. Yan, J.; Zhang, X.; Lei, Z.; Yi, D.; Li, S. Z. *Structural Models for Face Detection,* In Automatic Face and Gesture Recognition (FG), 10th IEEE International Conference and Workshops, 2013, pp 1–6.
149. Taigman, Y.; Yang, M.; Ranzato, M. A.; Wolf, L. In *Deepface: Closing the Gap to Human-level Performance in Face Verification,* Proceedings of the IEEE Conference on Computer Vision and Pattern Recognition, 2014, pp 1701–1708.
150. Li, H.; Lin, Z.; Shen, X.; Brandt, J.; Hua, G. In *A Convolutional Neural Network Cascade for Face Detection,* Proceedings of the IEEE Conference on Computer Vision and Pattern Recognition, pp 5325–5334.
151. Schroff, F., Kalenichenko, D., & Philbin, J. *Facenet: A unified embedding for face recognition and clustering,* In Proceedings of the IEEE Conference on Computer Vision and Pattern Recognition, 2015, pp 815–823.
152. Chowdhary, C. L.; Acharjya, D. P. Singular Value Decomposition–Principal Component Analysis-Based Object Recognition Approach. Bio-Inspired Comput. Image Video Process. 2018, 323.
153. Chowdhary, C. L. *Application of Object Recognition with Shape-Index Identification and 2D Scale Invariant Feature Transform for Key-Point Detection,* Feature Dimension Reduction for Content-Based Image Identification, 2018, 218–231.
154. Chowdhary, C. L.; Muatjitjeja, K.; Jat, D. S. *Three-Dimensional Object Recognition Based Intelligence System for Identification,* Emerging Trends in Networks and Computer Communications (ETNCC), 2015.
155. Chowdhary, C. L.; Ranjan, A.; Jat, D. S. Categorical Database Information-theoretic Approach of Outlier Detection Model. *Annal. Comput. Sci. Series.* 14th Tome 2nd Fasc. **2016,** 29–36.
156. Chowdhary, C. L. *Linear Feature Extraction Techniques for Object Recognition: Study of PCA and ICA. J. Serbian Soc. Comput. Mech.* **2011,** *5* (1), 19–26.
157. Chowdhary, C. L.; Acharjya, D. P. Breast Cancer Detection Using Hybrid Computational Intelligence Techniques. *Handbook of Research on Emerging Perspectives on Healthcare Information Systems and Informatics;* IGI Global: USA, 2018, 251–280.
158. Chowdhary, C. L.; Acharjya, D. P. Segmentation of Mammograms Using a Novel Intuitionistic Possibilistic Fuzzy C-Mean Clustering Algorithm. *Nature Inspired Comput.* **2018,** 75–82.
159. Chowdhary, C. L.; Acharjya, D. P. Clustering Algorithm in Possibilistic Exponential Fuzzy c-Mean Segmenting Medical Images. *J. Biomimet. Biomat. Biomed. Eng.* **2017,** *30,* 12–23.
160. Chowdhary, C. L.; Acharjya, D. P. A Hybrid Scheme for Breast Cancer Detection Using Intuitionistic Fuzzy Rough Set Technique. *Biometrics: Concepts, Methodologies, Tools, and Applications,* **2016,** 1195–1219.
161. Das, T. K. Chowdhary, C. L. Implementation of Morphological Image Processing Algorithm using Mammograms. *J. Chem. Pharm. Sci.* **2016,** *10* (1), 439–441.
162. Chowdhary, C. L. A Review of Feature Extraction Application Areas in Medical Imaging. *Int. J. Pharm. Technol.* **2016,** *8* (3), 4501–4509.
163. Chowdhary, C. L. Acharjya, D. P. In *Breast Cancer Detection using Intuitionistic Fuzzy Histogram Hyperbolization and Possibilitic Fuzzy c-mean Clustering Algorithms with*

Texture Feature Based Classification on Mammography Images, Proceedings of the International Conference on Advances in Information Communication Technology & Computing, 2016, p 21.

164. Chowdhary, C. L.; Sai, G. V. K.; Acharjya, D. P. Decreasing False Assumption for Improved Breast Cancer Detection. *J. Sci. Arts* **2016,** *35* (2), 157–176.
165. Chowdhary, C. L.; Sai, G. V. K.; Acharjya, D. P. Decrease in False Assumption for Detection Using Digital Mammography. *Comput. Intell. Data Mining* **2015,** *2*, 325–333.

CHAPTER 6

An Overview of the Concept of Speaker Recognition

SINDHU RAJENDRAN[1*], MEGHAMADHURI VAKIL[1], PRAVEEN KUMAR GUPTA[2], LINGAYYA HIREMATH[2], S. NARENDRA KUMAR[2], and AJEET KUMAR SRIVASTAVA[2]

[1]*Department of Electronics and Communication, R. V. College of Engineering, Bangalore 560059, India*

[2]*Department of Biotechnology, R. V. College of Engineering, Bangalore 560059, India*

[*]*Corresponding author. E-mail: sindhur@rvce.edu.in*

ABSTRACT

For communication, speech is one of the natural forms. A person's voice contains various parameters that convey information such as emotions, gender, attitude, health, and identity. Determination of these parameters will help to further develop the technology into a reliable and consistent means of identification using speaker recognition. Speaker recognition technologies have wide application areas especially in authentication; surveillance and forensic speaker recognition. In addition, speaker recognition refers to the automated method of identifying or confirming the identity of an individual based on his/her voice. Speech recognition strips out the personal differences to detect the words. Speaker recognition typically disregards the language and meaning to detect the physical person behind the speech. Speech recognition is language-dependent, while Speaker recognition is independent of language. In essence, voice biometrics provides speaker recognition rather than speech recognition. The most accepted form of identification for a human is his/her speech signal. Principally the speaker recognition is the computing task of validating a user's claimed identity using characteristics extracted from their voice. The speaker recognition process based on a speech signal is treated as one of the most exciting technologies of human recognition. For Speaker identification activities we mainly emphasize the physical features of signal. Speakers could be categorized as speaker identification and speaker verification. In speaker identification, the obtained features are compared with all the speaker's features which are stored in a voice model database and in speaker verification the obtained features are only compared with the stored features of the speaker he/she claimed to be. In this chapter the general principles of speaker recognition, methodology, and applications are discussed.

6.1 INTRODUCTION

Voice is one of the most authentic and accepted form of identification for humans. As specified earlier people voice contained distinctive determinations that gave data, for example, sentiments, sexual orientation, disposition, wellbeing, and character. Identification of these parameters is done using speaker recognition. Speaker recognition is also called as voice recognition system. Audio signal feature is classified either in perceptual mode or in physical mode. We mainly deal with the physical feature of speaker recognition. Speaker can be considered as speaker distinguishing proof and speaker confirmation. Speaker identification is identifying who the person is, whereas speaker verification is verifying whether the person spoken is authentic or not.[1] In speaker identification, the received specifications are noted and estimated with all the speaker's particulars which are contained in a voice show information gatherer and in speaker confirmation they got information are just contrasted and the put away highlights of the speaker he/she guaranteed to be.[2]

Every speaker recognition framework has two phases: enrollment and confirmation. At enrollment phase, different features are extracted from the speakers recorded voice. At the confirmation phase, the speaker's voice compared with formerly made voice prints which are present in the databases. For recognizable proof frameworks, the example is looked at against various voice models keeping in mind the end goal to choose the best match (es) while confirmation stage frameworks analyze an example against a solitary voice print. In view of the procedure included, confirmation is speedier contrasted with the recognizable proof. There are two kinds of speaker recognition to be specific dependent and free.[3] Content free frameworks are regularly utilized for speaker recognizable proof particularly in scientific applications as they require almost no participation by the speaker. For this state the content amid enrollment and test is diverse not at all like text-subordinate frameworks in which the content must be the same for enrollment and check. In a content ward framework, prompts can either be same over all speakers (normal expression) or one of a kind.

In speaker recognition system, the two important modules are feature extraction and feature matching. In this module, we mainly look into the block diagram of speaker recognition system, algorithms associated with feature extraction unit, algorithms associated with feature classification, and a huge number of applications of speaker recognition system.

6.1.1 SPEECH VERSUS SPEAKER

There is always confusion between speech and speaker recognition. Speech is nothing but interpretation of what is said and speaker recognition means identification of who has said. Speaker recognition mainly deals with the person who has uttered the word; here we mainly deal with the recognition of a person uniquely.[4] Both speech and speaker recognition have applications in different areas according to the interest.

6.1.2 BLOCK DIAGRAM

Speaker recognition has two parts based on the area of application, first is speaker identification and second one is speaker verification. In identification of speaker, speaker is identified based on his voice. In speaker verification, the voice is compared with the different databases which are stored earlier. The speaker identification has two parts, text dependent and another is text independent. In text dependent speaker identification, the same voice samples are taken for training as well as in test phase that means the same text is spoken at training and testing.[5,6] In text independent, voice samples may be different at training and testing phase. Feature extraction and feature classification are the two important parts of speaker identification. Figure 6.1 shows the block diagram of speaker recognition framework. At first, the info discourse is bolstered to the pre preparing unit to expel the quiet time frame from the discourse flag and then to the feature extractor unit wherein the necessary feature extraction algorithm using linear predictive coefficients (LPC), mel-frequency cepstral coefficients (MFCC), delta mel-frequency cepstrum coefficients (ΔMFCC), perceptual linear prediction coefficients (PLPC), RelAtive SpecTrAl

(Rasta)—PLPCC, linear predictive cepstral coefficients (LPCC) and wavelet transform (WT) are used to obtain the necessary specifications from the signal and finally classifying the signal using different classifiers such as support vector machine, vector quantization, dynamic time warping, and K-nearest neighbor algorithms.

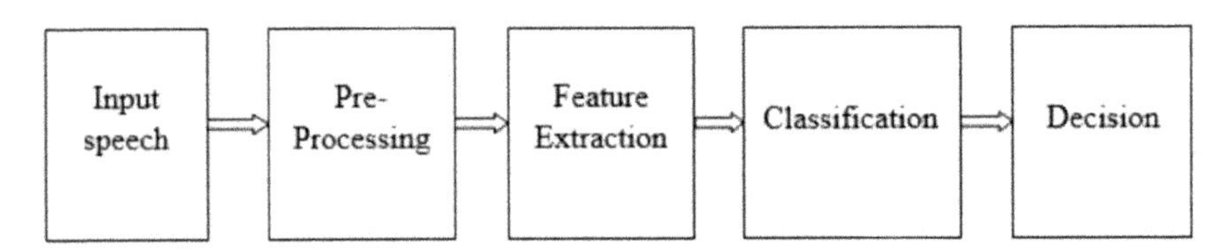

FIGURE 6.1 Block diagram of speaker recognition framework.
Source: Adapted from Ref. [23].

6.2 FEATURE EXTRACTION IN SPEAKER RECOGNITION

Feature extraction is an important part in identification of speaker. A raw data is extracted from the speakers' speech signal. Unwanted signal such as noise is removed by using various noise removal techniques and feature vector is constructed from the extracted useful information from speech signal. Before extracting features from the speech signal it is important to frame the signal and apply windowing on the frames, since the speech signal is varying continuously on large scale.[7] Framing the signal in small scale reduces the variation in speech signal and one can extract specific character of a speaker that is needed for identification. Windowing is applied on each frame to smooth the signal.

There are many techniques in feature extraction to extract features from the speech signal. Few techniques are explained in this section such as LPC.

6.2.1 LINEAR PREDICTIVE COEFFICIENT

Linear prediction coding extracts the features which represents the vocal tract resonance of speech signal.[8] The linear predictive filter predicts the present value based on the linear combination of previous sample values, hence the name linear prediction coefficient. It is represented as follows:

$$K(n) = a_1(K(n-1) + a_2(K(n-2) + a_3(K(n-3) + \ldots + a_m(K(n-m),$$

where $K(n)$ is present value, and $a_1 \ldots a_m$ are linear coefficients. LPC method is used to encode the speech signal at very low bit rate.

6.2.2 MEL FREQUENCY CEPSTRUM COEFFICIENTS

This is one of the popular feature extraction methods in of recognition of speaker. The process starts with framing the incoming speech signal (usually the size of the frame is power of two) and applying windowing to the framed signal. Figure 6.2 shows the block diagram of MFCC technique.[9]

The frequency components of sound which human ear perceives will not follow linear scale. Hence a Mel-scale is derived in order to mimic human ear, in which the scale is linear up to 1000 Hz and it is logarithmic above 1000 Hz.

Fourier transform is applied to signal after windowing to transform the signal from time to frequency domain. The equation is given by,

$$X(f) = \int_{-\infty}^{+\infty} x(t)\,\mathrm{e}^{-j2\pi ft}\,\mathrm{d}t.$$

The power spectrum of the Fourier transform is passed through filter bank whose frequencies are non-uniformly spaced. These frequencies are called Mel-frequencies

$$Mel(f_m) = {1000}\Big/{\ln(1+{10}/{7}) \times \ln\left(1+{f_m}/{700}\right)}.$$

The above equation shows logarithmic distribution of the Mel-frequencies, where f is the actual frequency in Hz. logs of the powers at each of the Mel frequencies is calculated and discrete Fourier transform of the Mel-frequencies log power is computed, whose spectrum amplitude are the MFCC.

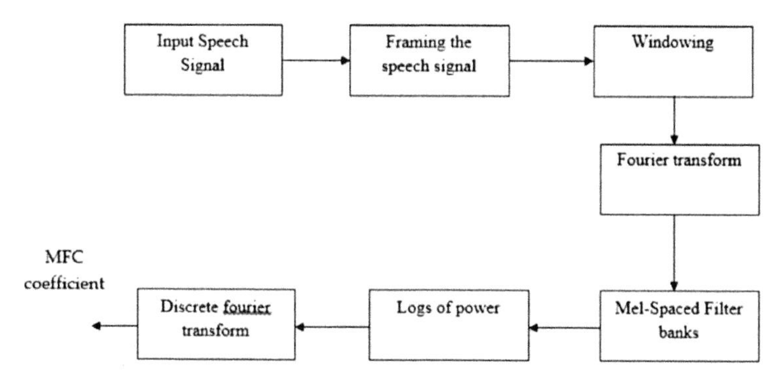

FIGURE 6.2 Block diagram of MFCC technique.

Source: Adapted from Ref. [23].

6.2.3 DELTA MEL-FREQUENCY CEPSTRUM COEFFICIENTS

One of the drawback of MFCC is it reflects the static characteristics but human ear is sensitive to static as well as dynamic signal characteristics.[10] Hence an improved version of MFCC is developed which is called as ΔMFCC. The ΔMFCC is appended to the first order of the MFCC to reproduce the dynamic characteristics of the signal. The equation is given as follows:

$$\Delta MFCC(q,r) = \frac{1}{\sqrt{\sum_{-k}^{k} j^2}} \sum_{j=-k}^{k} j \times MFCC(q, r+j).$$

6.2.4 PERCEPTUAL LINEAR PREDICTION COEFFICIENTS

This method is similar to LPC. In LPC method, linear prediction of speech signal is done by autoregressive moving average method whereas in PLPC, the perceptual processing of signal is done before the autoregressive process.[7,11] Figure 6.3 shows block diagram of PLPC process. PLPC has mainly three perceptual, first the critical band analysis, second equal loudness curve, and third one is intensity loudness power law.

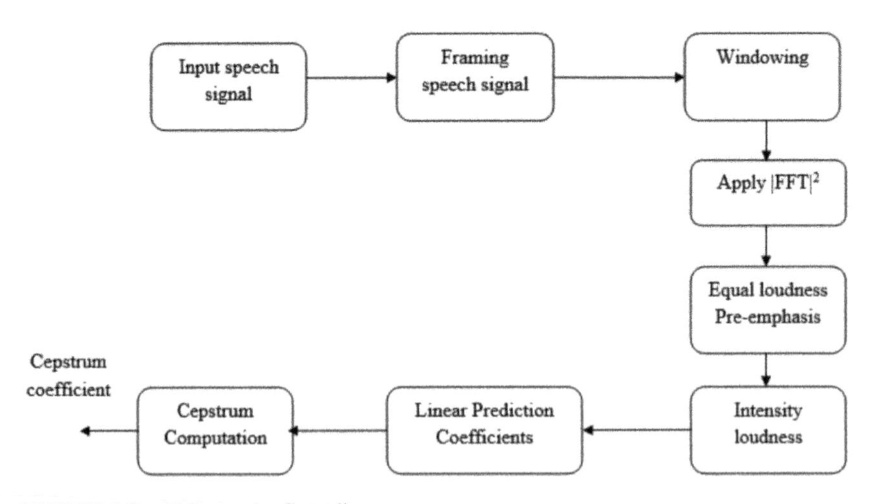

FIGURE 6.3 PLP process flow diagram.

The framed signal is applied to windowing technique. The windowed signal Fast Fourier Transform is calculated and the power spectrum is taken for next process and it is given as follows:

$$P(w) = Re(S(w))^2 + Im(S(w))^2.$$

Frequency warping is carried out by applying bark scale to the power spectrum. In order to match the human hearing resolution, the frequency is converted to bark. The bark frequency is calculated as follows:

$$\Omega(w) = 6\ln\left[\frac{w}{1200\pi} + \left[\left(\frac{w}{1200\pi}\right)^2 + 1\right]^{0.5}\right].$$

Convolution of frequency warped signal and the critical ban masking curves power spectrum is calculated to simulate the critical band integration of human hearing system. The convoluted signal is down sampled at an interval which is equal to one Bark. All the process is now integrated into Bark filter bank. The sensitivity of the human hearing system is simulated by applying filter-bank outputs to equal loudness pre-emphasis. The values equalized are now transformed by applying power law. The equalized value is raised to power 0.33. Linear prediction is applied to the line spectrum of auditory warped signal and the LP coefficients are computed. At last, by applying recursion to the LP coefficients, LP cepstral coefficients are obtained.

6.2.5 RELATIVE SPECTRAL (RASTA)—PLPCC

Usually Rasta method and PLPCC method are combined to get the low pass transfer function. In general, the rate of change of non-linguistic elements in speech is often present outside the rate of change of vocal tract.[12] The element which is changing very gradually or very rapidly than rate of change of speech is suppressed by the Rasta method. Rasta filtering method gives better accuracy in the presence of noise.

6.2.6 WAVELET TRANSFORM

Wavelet transform is simple and it simplifies the feature extraction process. The extraction of feature takes place in two bands; hence this technique is called as sub-band coding. The speech signal is divided into low frequency band and high frequency band.[13] Usually, high frequency band contains more of noise and the low frequency band contains the useful information of the signal.

Most of the signal energy is concentrated in the low frequency band. Hence the further level of decomposition starts from low frequency band.

6.2.7 LINEAR PREDICTIVE CEPSTRAL COEFFICIENTS

LPCC technique is an extension of the LPC technique. The LPCC is achieved by all pole model and LPCC coefficients are computed by the recursion from the LP coefficients to the LPC Cepstrum.[8,14] The coefficients in LPCC are represented in Cepstrum domain. Figure 6.4 shows the block diagram of LPCC.

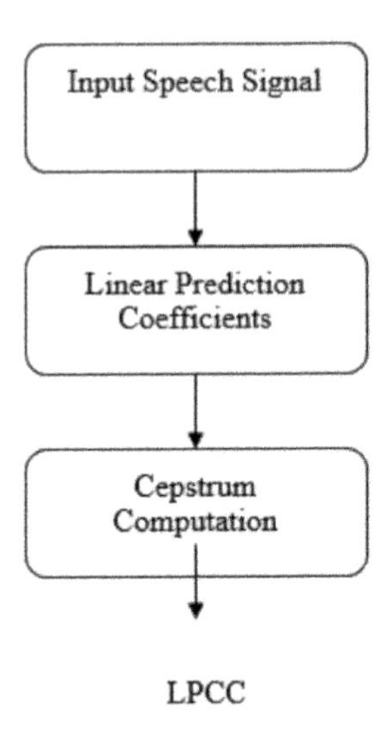

FIGURE 6.4 Block diagram of LPCC.

6.3 FEATURE CLASSIFICATION

Feature classification is the final stage in a speaker recognition system and an important part. The features patterns are classified into different classes. Some of the most widely used algorithms are SVM, DWT, VQ, GMM, and *k*-NN is discussed in detail in the next section. In this stage the features extracted are tallied with reference model and the choice is made. Many challenges are incurred in this stage since the class refers to individual speaker. Since here we are just managing characterization methodology in view of separated highlights, it can likewise be condensed as highlight coordinating.[15] To include more, if there exists an arrangement of examples for which the comparing classes are as of now known, at that point the issue is lessened to regulated example acknowledgment. These examples are utilized as preparing set and grouping calculation is resolved for each class. The rest designs are then used to test whether the grouping calculation works appropriately or not; accumulation of these examples is alluded as the test set. In the test set if there exists an example for which no arrangement could be inferred, and after that the example is considered as unregistered client for the speaker recognizable proof process. The robustness of algorithm is determined by identifying the registered users correctly and discarding the unknown users.

6.3.1 k-NN CLASSIFIER

The (*k*-NN) k-closest neighbor's algorithm is a technique used for differentiation and regression. KNN is a non-parametric arrangement strategy and its essential lead is to anticipate protest esteems in light of *k* tests that are near each other in the Euclidean space having a place with a similar class.[16] The info involves *k* nearest preparing cases in the component space.

With the object assigned to the most common class among its *k* nearest neighbors, the KNN classifier classifies the object by the majority voting of its neighbor (where *k* is a positive integer, typically small). If $k = 1$, then the class is defined by assigning object to single nearest neighbor.

In the order arrange, *k* is a customer described unfaltering, and an unlabeled vector (an inquiry or test point) is assembled by doling out the check which is most unending among the *k* getting ready tests nearest to that request point. A normally used partition metric for perpetual variables is Euclidean detachment. KNN can be upgraded basically if the partition metric is discovered with specific estimations, for instance, large margin nearest neighbor or neighborhood portions examination. A detriment of the basic bigger part voting course of action happens when the class dispersal is skewed. That is, instances of a more ceaseless class tend to manage the estimate of the new delineation, since they tend to be typical among the *k* nearest neighbors due to their broad number. One way to deal with vanquish this issue is to weight the gathering, considering the partition from the test point to each one of its *k* nearest neighbors.[17]

The most appropriate choice of *k* relies on the information. In the event that *k* is too little, at that point the outcome can be sensitive to noise. In the event that k is too enormous, at that point the outcome can be erroneous where neighbors incorporate an excessive number of focuses from different classes. For the most modules, larger estimations of *k* reduce the effect of noise on the differentiation; however, make boundaries between classes less particular.

The main disadvantage of KNN is the huge memory needed to store the whole training set. If the training set is large, response time will be also large which resulted in bad run-time performance. Moreover, KNN is affected by non-relevant or redundant attributes or features vectors which effect on the classification accuracy. Thus, the feature vector should be processed in advance and normalized.

Figure 6.5 shows the flow diagram of KNN.

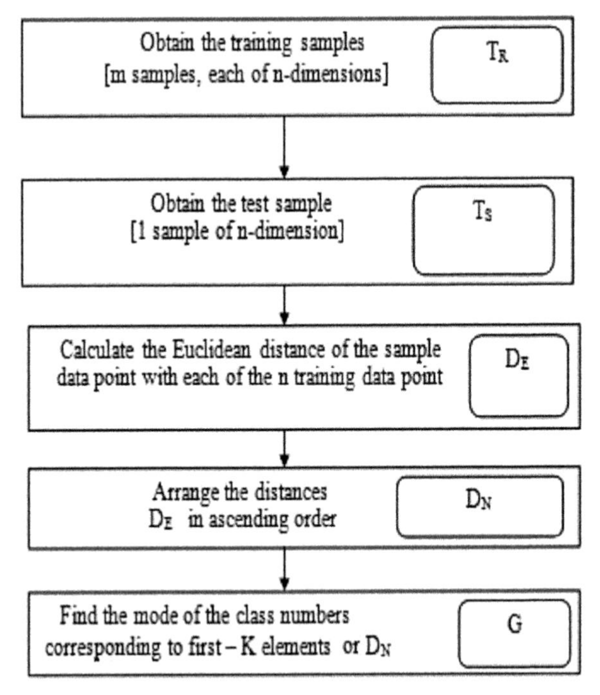

FIGURE 6.5 KNN diagram. DE, Euclidean distance; DN, Euclidean distances in ascending order; G, KNN result; TS, test data; and TR, training data.

In this case, *K* equals to one. Last row of the output of PCA corresponds to the test data TS. Rest of rows of PCA correspond to the training data matrix TR. Euclidean distance is computed using distance formula between the test data and training data. The test data sample is considered as the group corresponding to the smallest distance.

6.3.2 SUPPORT VECTOR MACHINE

Support vector machine (SVM) is mostly used for binary classification for speaker recognition. It is a simple, effective and supervised learning algorithm that analyzes datasets. Basically the algorithm obtains a set of input information and foretells for each given input, which of two likely classes form the output, by making it not depend on the probability binary linear classifier. Provided a stack of preparing datasets, each considered as belonging to one of two categories, an SVM training algorithm builds a model that assigns new examples into one category or the other. This technique is used for non-linear as well as for linear classification.[18,22–25] By technique called as kernel trick, SVMs can efficiently perform a non-linear classification, implicitly mapping inputs into high-dimensional feature spaces. An unsupervised approach is needed, when data is not labeled, this unsupervised approach forms a group of natural clustering of data, to these newly formed group anew data is mapped. The method using clustering algorithm is used in many industrial applications. Figure 6.6 shows the flow diagram of SVM classifier.

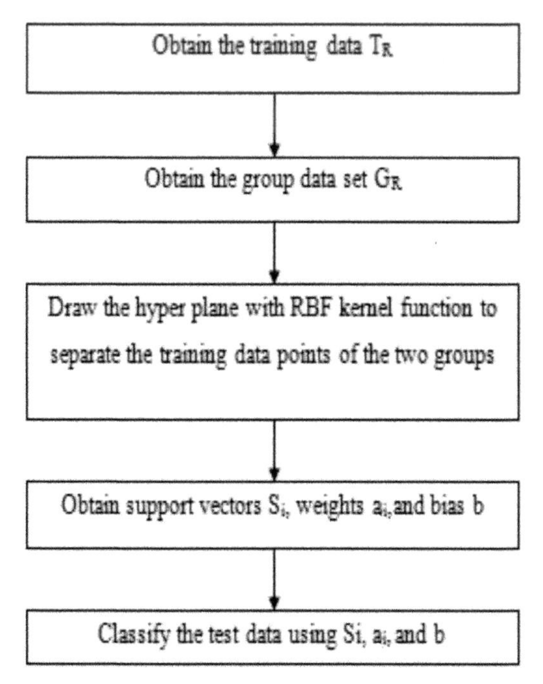

FIGURE 6.6 SVM flow diagram. A_i, weights; B, bias; G, original target group data; T_S, test signal; S_i, support vectors; and T_R, training data matrix.

The binary SVM works in two phases namely the train phase and the test phase. The input to the algorithm is the train data matrix corresponding to two classes. The target group data consists of the required classifier output to each of the train signals. The output of the classifier is class 1 or class 2, as corresponding to the test signal.

In the train phase a hyper plane is drawn according to the kernel function. Support vectors S_i, weights A_i, and bias is calculated. These are used to classify the test signal between class 1 and class 2 in test phase.

6.3.3 DYNAMIC TIME WARPING

Dynamic time warping algorithm calculates the separation between two successions which may fluctuate in time or speed. Time warping is done non-linearly to normalize the timing differences between test utterance and the reference template. Then time normalization distance is calculated between patterns. Authentic speaker is identified with minimum time normalized distance.[19,22] It is beneficial for variable length input feature and require less storage space. This algorithm is connected to fleeting arrangements of sound and video that is any information that can be transformed into a direct grouping can be broke down.

The algorithm works with certain rules and restrictions:

1. Every file from the primary arrangement must be coordinated with at least one file from the other succession, and the other way around.
2. The first record from the main grouping must be coordinated with the principal list from the other arrangement (however, it does not need to be its solitary match).
3. The last list from the principal grouping must be coordinated with the last list from the other succession (yet it does not need to be its solitary match).
4. The mapping of the records from the principal grouping to files from the other arrangement must be monotonically expanding, and the other way around, that is, if are files from the main succession, at that point there must not be two lists in the other succession, to such an extent that file is coordinated with file and file is coordinated with file, and the other way around.

The choice is taken by the match that fulfills every one of the confinements and the standards and which has the insignificant cost. The successions are "twisted" in the time measurement non-straightly to decide a measure of their likeness free of certain non-direct varieties.

Notwithstanding a comparability measure between the two groupings, a "distorting way" is created, by twisting as indicated by this way the two signs might be adjusted in time. DWT discovers applications in hereditary arrangement and sound synchronization.

6.3.4 VECTOR QUANTIZATION

Vector quantization is classical quantization technique from signal processing. It gives us the degree of match between the test data and reference model. It allows modeling of probability density functions by distribution of prototype vector. By using vector quantization, the extracted speech features of speaker which are of large set are quantized to a number of centroids. These centroids compose the codebook of that speaker.[20] It is used for data compression and requires less storage. VQ is computationally less complex. Memory requirement is achievable for real time applications.[21] Vector quantization has applications in density estimation and loss data compression because of its low error rate.

6.3.5 GAUSSIAN MIXTURE MODEL

The other technique for classifier is Gaussian mixture model (GMM) which is a density estimator. It is an unsupervised learning algorithm which requires less training and test data thereby giving better performance. GMM requires less memory and fewer amounts of data to train the classifier.[22] Expectation maximization algorithm is used to estimate GMM parameter from training data. A sequence of feature extracted from input signal. By computing log likelihood, the distance of the given sequence from the model is obtained.

6.4 APPLICATIONS

Speaker recognition has wide areas of applications especially in areas such as authentication, surveillance, and forensic (Fig. 6.7).

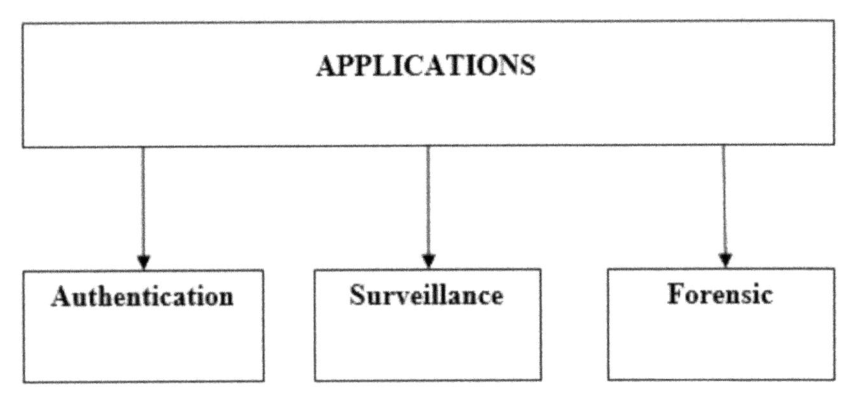

FIGURE 6.7 Applications.

1. In authentication users identify using person's voice. It is also known as biometric person authentication. It is less erroneous compared to other authentication processes because each person has unique voice and it has unique physiology, anatomy, and learned habits that a person uses in his everyday life to recognize the person.
2. High quantities of data in electronic eavesdropping of telephone and radio conversations, filter mechanisms must be applied for authentic information. One of such filters may be recognition of speakers that are of interest.
3. During the crime if there is a speech sample recorded it can be compared with the suspect's voice to give an identification of the similarity of two voices.

There are more wide areas of application in security, speech recognition, multispeaker tracking, and personalized user interfaces.

1. Along with face recognition, speaker recognition is also used in credit card transactions as an authentication method. Speaker recognition is an application of any biometric authentication technique. By this technology, transaction authentication facility or computer access control, voice authentication for long distance calling.
2. To reduce the speaker variability, speaker recognition technology can be used in speech recognition systems by speaker adaption by making use of speaker gating unit which recognizes who is speaking thereby adapting its speech recognizer parameters for better performance of current speaker.

3. Multispeaker tracking is observed especially in teleconference when there is no familiarity between attendants. There are three types of multispeaker tasks are speaker detection, speaker tracking, and speaker segmentation. Speaker detection is determining whether known speaker is present. Speaker's speaking intervals are located in the recordings. Speaker segmentation consists of locating speech intervals of different speakers.
4. Personalized user interface such as voice mail in general are popular because of development in speech technology, the system adapts to the needs and preferences of person by recognizing the speaker. Therefore, the requirement of speaker recognition technique is involved in telephone aided services under different kinds of acoustic conditions such as in office, street, and so on.

6.5 CONCLUSIONS

For the communication, speech is one of the natural forms. A person's voice contains various parameters that convey information such as emotions, gender, attitude, health, and identity. Determination of these parameters will help to further develop technology into a reliable and consistent means of identification using speaker recognition. Speaker recognition technologies have wide application areas especially in the authentication; surveillance, and forensic speaker recognition. In addition, speaker recognition refers to the automated method of identifying or confirming the identity of an individual based on his/her voice. Speech recognition strips out the personal differences to detect the words. Speaker recognition typically disregards the language and meaning to detect the physical person behind the speech. Speech recognition is language dependent, while Speaker recognition is independent of language. In essence, voice biometrics provides speaker recognition rather than speech recognition. The most accepted form of identification for human is his/her speech signal. Principally the speaker recognition is the computing task of validating a user's claimed identity using characteristics extracted from their voice. The speaker recognition process based on a speech signal is treated as one of the most exciting technologies of human recognition. For Speaker identification activities, we mainly emphasize on the physical features of signal. Speaker could be categorized as speaker identification and speaker verification. In speaker identification, the obtained features are compared with all the speaker's features which are stored in a voice model database and in speaker verification the obtained features are only compared with the stored features of the speaker he/she claimed to be. In this chapter, the general principles of speaker recognition, methodology, and applications are discussed.

In this chapter, an overview on the concept of speaker recognition includes the necessity and importance of speech, a brief introduction to speaker recognition using the block diagram is explained. The principle difference between the speaker and speech recognition is been discussed and the whole methodology and the algorithms for feature extraction process and algorithms for classifiers. Due to the importance of speaker recognition it has wide applications in different areas, some of the main applications in authentication, surveillance and forensic are also discussed.

KEYWORDS

- **parameters**
- **speaker recognition**
- **speaker identification**
- **speaker verification**
- **voice**

REFERENCES

1. Mikhael, W. B.; Premakanthan, P. *Speaker Identification Employing Redundant Vector Quantisers*; Electronics Letters, 2002; pp 1396–1398.
2. Premakanthan, P.; Mikhael, W. B. In*Speaker Verification/recognition and the Importance of Selective Feature Extraction: Review*. Proceedings of the 44th IEEE Midwest Symposium (MWSCAS) on Circuits and Systems, 2001; pp 57–61.
3. Ozaydin, S. In *Design of a Text Independent Speaker Recognition System*, International Conference on Electrical and Computing Technologies and Applications (ICECTA), 2017; pp 1–5.
4. Tang, Z.; Li, L.; Wang, D. In *Multi-task Recurrent Model for Speech and Speaker Recognition*, Asia-Pacific Signal and Information Processing Association Annual Summit and Conference (APSIPA), 2016; pp 1–4.
5. Tirumala, S. S.; Shahamiri, S. R.; Garhwal, A. S.; Wang, R. Speaker Identification Features Extraction Methods: a Systematic Review. *Expert Syst. Appl.* **2017,** *90* (30), 250–271.

6. Marciniak, T.; Weychan, R.; Stankiewicz, A. A.; Dabrowski, A. Biometric Speech Signal Processing in a System With Digital Signal Processor. *Bull. Polish Acad. Sci. Tech. Sci.* **2014,** *62* (3), 589–594.
7. Chowdhary, C. L.; Acharjya, D. P. In *Breast Cancer Detection Using Intuitionistic Fuzzy Histogram Hyperbolization and Possibilitic Fuzzy c-mean Clustering algorithms With Texture Feature-based Classification on Mammography Images*. Proceedings of the International Conference on Advances in Information Communication Technology & Computing, 2016; p 21.
8. Sunny, S.; Peter, S. D.; Jacob, K. P. In *Feature Extraction Methods Based on Linear Predictive Coding and Wavelet Packet Decomposition for Recognizing Spoken Words in Malayalam*. International Conference on Advances in Computing and Communications, 2012; pp 27–30.
9. Chauhan, P. M.; Nikita P. Desai. In *Mel Frequency Cepstral Coefficients (MFCC) Based Speaker Identification in Noisy Environment Using Wiener Filter*, International Conference on Green Computing Communication and Electrical Engineering (ICGCCEE), 2014; pp 1–5.
10. Wu, J.; Yu, J. In *An Improved Arithmetic of MFCC in Speech Recognition System*, International Conference on Electronics, Communications and Control, (ICECC), 2011; pp 719–722.
11. Alam, M. J.; Kinnunen, T.; Kenny, P.; Ouellet, P.; O'Shaughnessy, D. Multitaper MFCC and PLP Features for Speaker Verification Using i-vectors. *Speech Commun.* **2013,** *55* (2), 237–251.
12. Nayana, P. K.; Mathew, D.; Thomas, A. In *Performance Comparison of Speaker Recognition Systems Using GMM and I-Vector Methods With PNCC and RASTA PLP Features*. International Conference on Intelligent Computing, Instrumentation and Control Technologies (ICICICT), 2017; pp 438–443.
13. Daqrouq, K.; Al-Qawasmi, A.-R.; Al-Sawalmeh, W.; Hilal, T. In *Wavelet Transform Based Multistage Speaker Feature Tracking Identification System Using Linear Prediction Coefficient*. International Conference on Advances in Computational Tools for Engineering Applications, 2009; pp 173–179.
14. Zhu, J.; Liu, Z. In *Analysis of Hybrid Feature Research Based on Extraction LPCC and MFCC*. 10th International Conference on Computational Intelligence and Security, 2014; pp 732–735.
15. Paulose, S.; Mathew, D.; Thomas, A. In *A Comparative Study of Text-Independent Speaker Recognition Systems Using Gaussian Mixture Modeling and I-Vector Methods*. International Conference on Intelligent Computing, Instrumentation and Control Technologies (ICICICT), 2017; pp 444–448.
16. Kacur, J.; Vargic, R.; Mulinka, P. In *Speaker Identification by K-Nearest Neighbors: Application of PCA and LDA Prior to KNN*. International Conference on Systems, Signals and Image Processing, 2011; pp 1–4.
17. Kacur, J. In *Modifications of KNN Classifier for Speaker Identification System*. IEEE International Symposium ELMAR, 2016; pp 35–38.
18. Chakroun, R.; Zouari, L. B.; Frikha, M.; Hamida, A. B. In *A Hybrid System Based on GMM-SVM for Speaker Identification*. IEEE Recent Advances in Intelligent Computational Systems (RAICS), 2015; pp 139–144.

19. Kau, K.; Jain, N. Feature Extraction and Classification for Automatic Speaker Recognition System—A Review. *Int. J. Adv. Res. Comput. Sci. Softw. Eng.* **2015,** *5* (1) 1–6.
20. Tiwari, V. MFCC and its Applications in Speaker Recognition. *IEEE Int. J. Emerg. Technol.* **2010,** *1* (1), 19–22
21. Farah, S.; Shamim, A. In *Speaker Recognition System Using Mel-Frequency Cepstrum Coefficients, Linear Prediction Coding and Vector Quantization.* 3rd IEEE International Conference on Computer, Control and Communication (IC4), 2013; pp 1–5.
22. Chowdhary, C. L.; Acharjya, D. P. Singular Value Decomposition–Principal Component Analysis-Based Object Recognition Approach. *Bio-Insp. Comput. Image Video Process.* **2018,** *1,* 323–341.
23. Ramgire, J.B.; Jagdale, S.M. A Survey on Speaker Recognition With Various Feature Extraction And Classification Techniques. *Int. Res. J. Eng. Tech.* **2016,** 3 (4).

CHAPTER 7

Analysis of Unimodal and Multimodal Biometric System

CHIRANJI LAL CHOWDHARY*

Department of Software and System Engineering,
School of Information Technology and Engineering (SITE),
Vellore Institute of Technology, Vellore, 632014 Tamil Nadu, India

**Corresponding author. E-mail: c.l.chowdhary@gmail.com*

ABSTRACT

In everyday life, the demand of biometric system is increasing. The study of this chapter is focused on the several kind of biometric modalities. A human everyday life science coordination naturally the whole thing for individual being identification. The science and engineering technology of motion and analyzing living information is mentioned as Biometric. The daily life systems are of such type that this may suggest a lot of security to handler. The biometric identifiers are regularly organized as physiological versus behavioral features. Physiological traits are acknowledged with the physical state-owned of the body. We are discussing here about several types of biometric classifications, that is, unimodal biometric systems and multimodal biometric systems. Throughout this analysis work mainly deliberate the unimodal and multimodal based work.

7.1 WHAT IS UNIMODAL AND MULTIMODAL BIOMETRIC SYSTEM?

Biometric technology depends on the proposition of measuring and examining the biological traits of a person and extracting the particular features out of the collected data and then using it for comparisons with other templates stored in the database. These unique biological traits are often known as biometric identifiers and can be categorized into mainly two types. The first one being the physiological identifier and the second one being behavioral identifiers. The physiological identifiers are associated with the shapes and structure of the body, on the other hand, the behavioral identifiers are associated with behavioral patterns of an individual. The physiological biometric identifiers consist of these following traits: fingerprints, face, iris, palm, hand, geometry, etc., whereas the behavioral consists of characteristics or behaviors, such as voice, gait, typing rhythm, etc.

In today's world, many different forms of access control have emerged, which includes token-based identification systems, such as driver's license or passport, and knowledge-based identification systems, such as a password or personal identification number. Even though these access controls are available biometric identifiers serve as highly reliable and secure than the token and knowledge-based methods but the one limitation is that the collection of biometric identifiers could increase the privacy concerns of the information being used.

7.2 THE REASONS FOR CHOOSING A BIOMETRIC TECHNOLOGY

Authentication, authorization, and accountability constitute the main components of security in a system. Authentication stands as the main feature out of the three because anyone who wants to access the system needs to be authenticated initially before allowing a user to access the system. The conventional authorization systems in-use at present are passwords and smartcards.

One of the crucial factors for opting biometric technology in place of traditional identification techniques, such as username and password, is that traditional methods are very vulnerable to security breaches and do not furnish satisfactory security system in the connected networks and remoting and data-driven world. The traditional identifications are prone to theft and can be misused in order to disguise anyone.

Biometric stands apart from all these because it depends on the inherent characteristics of an individual. The authentication provided by biometric are much stronger and cannot be guessed, forgotten, misplace or effortlessly forged unlike the other techniques, such as possession-based and knowledge-based.

Before using a biometric trait in any application, these traits are to be assessed on majorly these seven of the following features. In order to provide high-performance and accuracy, these factors must be assessed in the biometric identification system.

1. Uniqueness: Biometric samples that are used to uniquely identify the subject.
2. Universality: This is an important feature every subject must possess this quality.
3. Permanence: Biometric traits should remain unaffected over a period of time.
4. Measurability: The biometric samples should be effortless to measure and acquire.
5. Performance: The process of acquiring or using must be robust.
6. Acceptability: The modality that is being used must be acceptable by all.
7. Circumvention: The vulnerability of the traits to get spoof attacks and identity frauds.

7.3 UNIMODAL BIOMETRIC SYSTEM

Unimodal biometric utilizes individual biometric attribute (either of one such as physical or biometric trait) to identify physiological life science adjust exhibit fingerprints, full of mathematics, eye and ear patterns, facial expressions, etc. Behavioral attributes exhibit features, such as voice, signatures, typewriting, and patterns, etc., on the other hand, in the process of recognizing a feature of a person, there are quits in selecting a real subject as an intruder or a fraud or a fraud as a real subject. For example, biometric systems support traits, such as iris or fingerprint or voice or gait, etc. (Fig. 7.1).

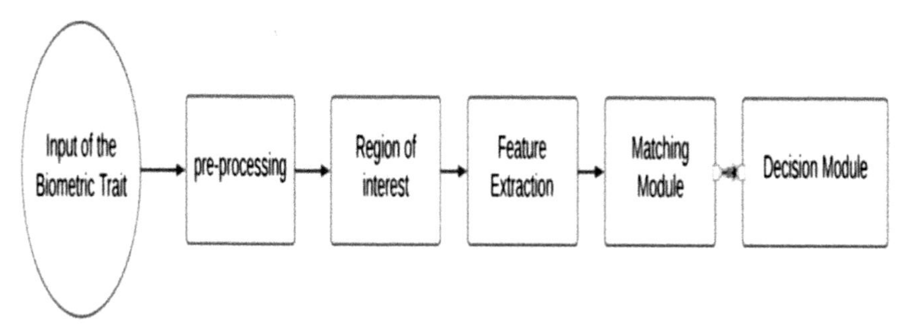

FIGURE 7.1 Simple block diagram of biometric system.

Iris recognition is the emerging branch in the field of biometric recognition. An iris resides in between pupil and sclera and is angular in shape. There are many distinct features while considering freckles, coronas, stripes, furrows, and so on.

Iris recognition is preferred by most of the people due to the following reasons:

- Uniqueness: The probability of a matching found in iris pattern is less than 0.35. Even though the iris comparison made between twins, their iris measure is abnormally different. This marks as a remarkable characteristic in the iris recognition system therefore widely used and is less prone to vulnerability.
- Reliability: Iris resides in the inner part of the eyes and is surrounded by the lid, lash, and tissue layer for the protection of the iris. It is unlike finger and palm that are much prone to damage and therefore the errors caused by recognition are very low and the quality of match produced accounts to 90% success.

The following are the primary steps that are followed in the iris recognition system:

- Segmentation: A method is required, such as to detach and omit the artifacts, in addition to locating the circular iris region therefore by segmentation we can calculate the inner and the outer boundaries of the iris area unit.
- Normalization: Capturing of iris of different people could be done using numerous sizes, for constant people additional size varies owing to the fluctuation in brightness and various other alternative factors. This method is capable of producing iris regions consisting of ditto consistent dimensions in order that beneath various conditions that the two photographic images of a common iris can consist of characteristic options at a constant geographical location.
- Feature extraction: The process is used to make comparisons between the templates, an iris pattern is taken and all the necessary features are encoded for further use. Several iris recognition systems in order to create a biometric model use bandpass decomposition of iris images. A feature vector is build that comprises the ordered sequence of options that are derived from the various interpretation of the iris images, these vectors are also known as copious texture.

- Matching of an image: To use "one-one" or "one to many" template match in order to verify via identification. A template that has been created by imaging the iris is evaluated and compared with a stored template that has been created and stored previously in a database. An identification is positive and precise if and only if the hamming distance is below the decision threshold. For example, hamming distance produces a result that is accurate and precise and results in a positive identification match.
- Localization: This is a vital step in the iris recognition process. A precise result is obtained only when one segment iris correctly form the original image. Localization of iris can be defined by detecting the location of iris' inner boundary and outer boundary.

7.4 FINGERPRINT RECOGNITION

It is the widely known and most practiced biometrics in actual execution. Fingerprint biometric identification has been used since centuries due to its uniqueness and consistency over the time used for identification purpose and also due to its enhancement in computing capabilities it has now become almost fully automated and there are no discrepancies. Due to the ease provided by the acquisition these fingerprint identifications have grown prominently as for the collection of fingerprints, there are numerous resources available and their entrenched utilization and acquiring by law enforcement and immigration.

7.4.1 FINGERPRINT PATTERNS

The three basic patterns of fingerprint ridges area are the arch, the loop, and the whorl.

An arch is formed wherever the ridge enters one aspect of the finger and then rises within the center forming an arch and exits on the opposite aspect of the finger.

Whereas with the loop the ridge enters one aspect of the finger and then a curve is formed and exits on an equivalent aspect of the finger forming a curve and exiting on an equivalent aspect of the finger from that it has entered.

Finally, a whorl is the pattern in which a ridge kind of circular pattern forms around a central purpose.

7.4.2 MINUTIAE FEATURES

Minutiae are area units that are very small or tiny details in a fingerprint that are most significantly recognized during the recognition of the fingerprint.

There are three major varieties of trivia features:

- The ridge ending, the bifurcation, and the spot.
- The ridge ending is a spot where the ridge ends. A bifurcation is that spot where there is a split found in the ridges and becomes two distinct ridges. Spots are units that are fingerprint ridges that have areas considerably shorter than other ridges.

7.5 LIMITATIONS OF UNIMODAL BIOMETRIC SYSTEMS

The successful establishment of biometric systems in different regular citizen applications does not suggest that biometrics is a completely tackled issue. Biometric systems that work by utilizing any single biometric trademark have the accompanying restrictions:

7.5.1 NOISE IN DETECTED INFORMATION

The detected information may be noisy or twisted. A fingerprint with a scar or a voice modified by cold is a case of noisy data. Noisy data could likewise be the aftereffect of defective or despicably kept up sensors (e.g., the collection of dirt on a fingerprint sensor) or unfavorable ambient conditions (e.g., a poor light of a client's face in a face recognition system). Noisy biometric information might be erroneously matched with templates in the database resulting in a client being incorrectly dismissed.

7.5.2 INTRACLASS VARIATIONS

The biometric information gained from a person during authentication might be altogether different from the information that was utilized to produce the template during enrollment, in this way affecting the matching procedure. This variation is commonly caused by a client who is mistakenly interacting with the sensor or when sensor attributes are altered (e.g., by changing sensors—the sensor interoperability issue) during the verification phase. As another case, the varying psychological makeup of an individual may result in endlessly different behavioral traits at different time instances.

7.5.3 DISTINCTIVENESS

While a biometric trait is required to vary essentially crosswise over people, there might be larger interclass similarities in the feature sets used to represent these traits. This confinement limits the discriminability gave by the biometric traits.

7.5.4 NONUNIVERSALITY

While each client is required to have the biometric trait being obtained, in reality, it is possible for a subset of the clients to not have a specific biometric. A unique mark biometric system, for instance, might be not able to extract features from the fingerprints of certain people, because of the poor quality of the ridges. Therefore, there is a failure to enroll (FTE) rate associated with utilizing a single biometric trait. It has been empirically assessed that as much as 4% of the population may have poor quality fingerprint ridges that are difficult to image with the currently available fingerprint sensors and result in FTE errors. Den Os et al.[1] report the FTE issue in a speech recognition system.

7.5.5 SPOOF ASSAULTS

An impostor may attempt to spoof the biometric trait of a legitimately enrolled client keeping in mind the end goal to circumvent the system. This kind of assault is particularly relevant when behavioral traits, for example, signature[3] and voice[2] are utilized. However, physical qualities are likewise vulnerable to spoof attack. For instance, it has been demonstrated that it is possible (although troublesome and cumbersome and requires the assistance of a legitimate user) to build artificial fingers/ fingerprints in a reasonable amount of time to circumvent a fingerprint verification system.[4]

7.6 MULTIMODAL BIOMETRIC SYSTEM

Multimodal biometric system (Fig. 7.2) utilizes two or more features or attributes of a person for authentication, that is, recognized together and are combined afterward to form a multimodal biometric system. In the aspect of improving the coverage of the population, deterring attacks caused by spoof attacks and increasing the degree of freedom and reducing the FTE. A multimodal biometric system can prominently improve the recognition performance and can lower the demand of the unimodal biometric system. The demand for storage requirement, processing time, and computational demand of biometric system can be less in unimodal as compared with multimodal biometric system.

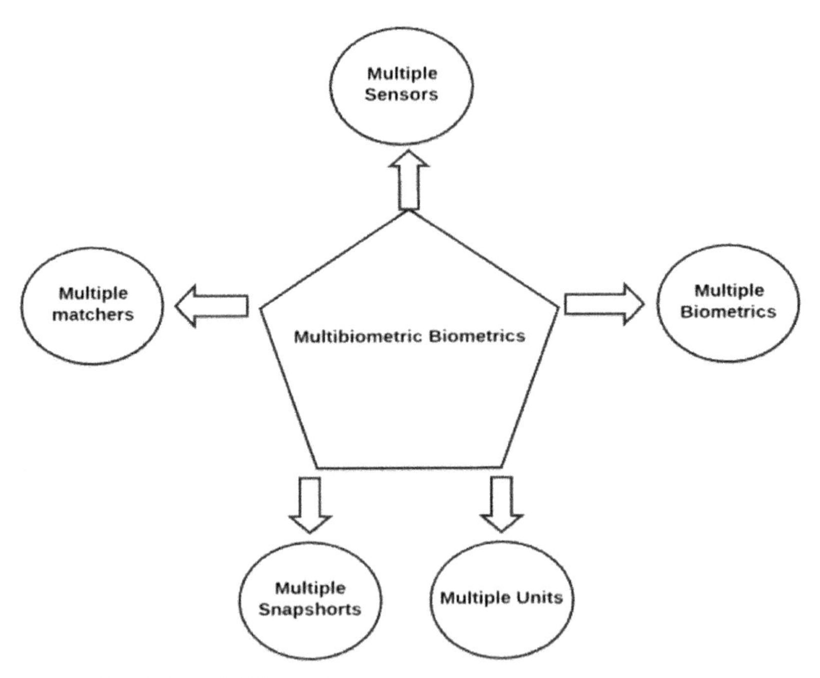

FIGURE 7.2 Multimodal biometric system.

7.7 WORKING OF MULTIMODAL BIOMETRIC SYSTEM

The following are the modules consisted by multimodal biometric system:

1. Sensor modules.
2. Feature extraction.

3. Matching module.
4. Decision-making module.

In multimodal biometric systems (Fig. 7.3) fusion is accomplished by performing at least two biometric characteristics against two different calculations which is then used to land at a choice. This sort of a procedure ends up being to a great degree valuable in circumstances, for example, an expansive scale civil ID situation, where the personality of thousands of individuals should be confirmed at one time. Likewise, having an extra technique for verification conquers the likelihood of inconvenience that can be caused by the failure of the essential biometric input.

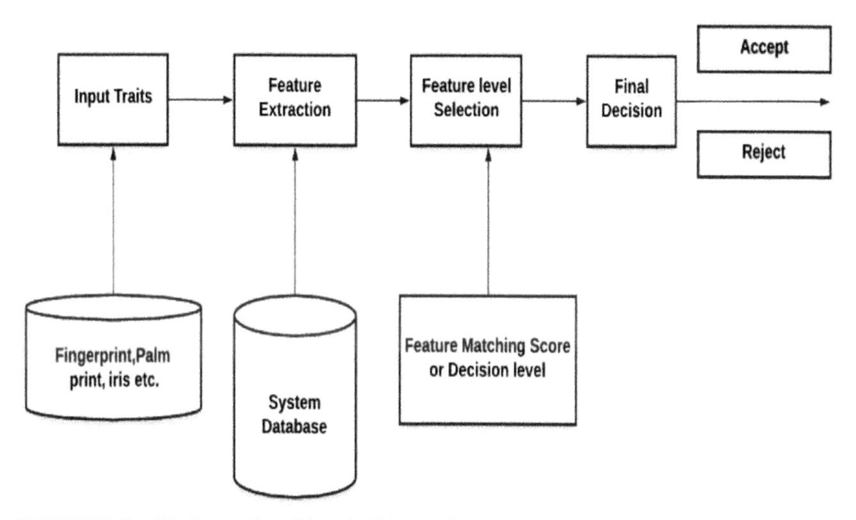

FIGURE 7.3 Working of multimodal biometric system.

7.8 HOW TO INTEGRATE?

Multimodal biometric systems can be intended to operate in one of the accompanying five situations:

7.8.1 MULTIPLE SENSORS

The information got from various sensors for the same biometric are joined. For example, optical, solid-state, and ultrasound-based sensors are available to catch fingerprints.

7.8.2 MULTIPLE BIOMETRICS

Multiple biometric characteristics, for example, fingerprint and face are joined. These systems will necessarily contain in excess of one sensor with every sensor detecting an alternate biometric characteristic. In a verification system, the various biometrics are normally used to enhance system exactness, whereas in an identification system the matching speed can likewise be enhanced with an appropriate combination scheme (e.g., face matching, which is commonly quick yet not extremely precise, can be utilized for recovering the best matches and afterward fingerprint matching, which is slower yet more exact, can be utilized for making the final identification decision).

7.8.3 MULTIPLE UNITS OF THE SAME BIOMETRIC

Fingerprints from at least two fingers of a man might be combined, or one image each of the two irises of a man might be consolidated.

7.8.4 MULTIPLE SNAPSHOTS OF THE SAME BIOMETRIC

More than one occurrence of the same biometric is utilized for enrollment and recognition. For instance, various impressions of the same finger, numerous sample of the voice, or multiple images of the face might be combined.

Multiple representations and matching algorithms for the same biometric, this includes combining diverse approaches to feature extraction and matching of the biometric characteristics. This could be utilized as a part of two cases. Initial, a verification or an identification system can utilize such a combination scheme to settle on a recognition decision. Second, an identification system may utilize such a combination scheme for indexing.

In situation 1, multiple sensors are utilized to detect the same biometric identifier while situation 2 utilizes numerous sensors to detect diverse biometric identifiers. A case of situation 1 might be the utilization of various cameras mounted to catch different perspectives of a man's face. A case of situation 2 is the utilization of a camera for catching face and an optical sensor to catch a unique finger impression. While situation 1 consolidates modestly independent information, situations 2 and 3 combine independent (or pitifully reliant) information and are required to bring about a substantially bigger change in recognition accuracy.

However, this improvement comes at the cost of inconvenience to the client in giving different signs and a more drawn out acquisition time. In scenario 4, just single information might be procured during recognition and matching with a few stored templates obtained during the one-time enrollment process; on the other hand, more information acquisitions might be set aside a few minutes of recognition and used to consolidate the coordinating against a single/multiple templates. Situation 5 joins distinctive representations and matching algorithms to enhance the recognition precision.

7.9 DIFFERENT TYPES OF BIOMETRIC SYSTEMS

7.9.1 MULTIALGORITHMIC BIOMETRIC SYSTEMS

These systems take a solitary biometric test from a solitary sensor and after that procedure it utilizing at least two different calculations.

7.9.2 MULTI-INSTANCE BIOMETRIC SYSTEMS

These systems utilize at least one sensor to catch tests of at least two different tests of the same biometric quality. A case of this could be a framework catching pictures of different dangers.

7.9.3 MULTISENSORIAL BIOMETRIC SYSTEMS

These systems utilize at least two distinctly different sensors to catch a similar occurrence of a biometric attribute. These caught tests are then prepared to utilize a single calculation or a mix of calculations. The case of multisensorial biometric systems is where a similar facial picture is caught utilizing a noticeable light camera and an infrared camera fixed with a specific frequency.

7.9.4 FUSION IN MULTIMODAL BIOMETRIC SYSTEMS

We are using multiple biometric modalities to make fusion of them to make a multimodal biometric system. This emerges the need to plan an instrument which can consolidate the classic cation result from each biometric channel and this component is known as a biometric fusion. This fusion combines the estimations from different biometric credits to improve the qualities and decline the shortcomings of the individual measurements (Fig. 7.4).

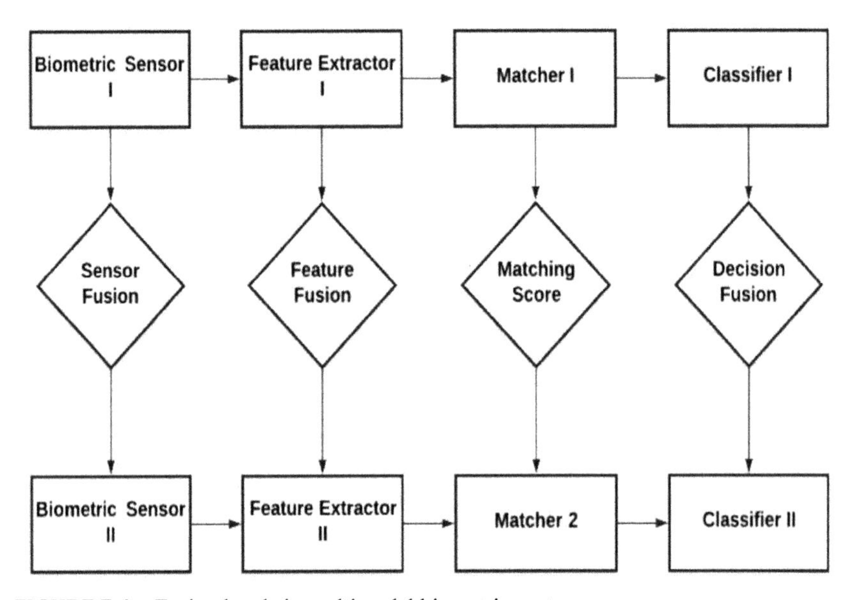

FIGURE 7.4 Fusion levels in multimodal biometric systems.

Fusion can be utilized to address various issues faced in the implementation of biometric systems, for example, precision, efficiency, vigor, applicability, and universality. There are different levels of intertwining the biometric characteristics that can be utilized to build heartiness of the multimodal biometric framework. They are sensor level fusion, feature level fusion, matching score level fusion and choice level combination.

7.9.5 SENSOR LEVEL FUSION

In sensor level fusion, we intertwine the biometric attributes originating from different sensors, for example, fingerprint scanner, an iris scanner, camcorder, and so on, to form a blended biometric characteristic and process.

7.9.6 FEATURE LEVEL FUSION

In feature level fusion, signals originating from different biometric channels are first prepared after and then the component vectors are extracted independently from each biometric characteristic. The feature vectors are then consolidated to form a composite feature vector utilizing a specific fusion classification and after that utilized for advanced classification. In feature level fusion, some diminishment methods should be utilized as a part of the request to choose just the helpful highlights (Fig. 7.5).

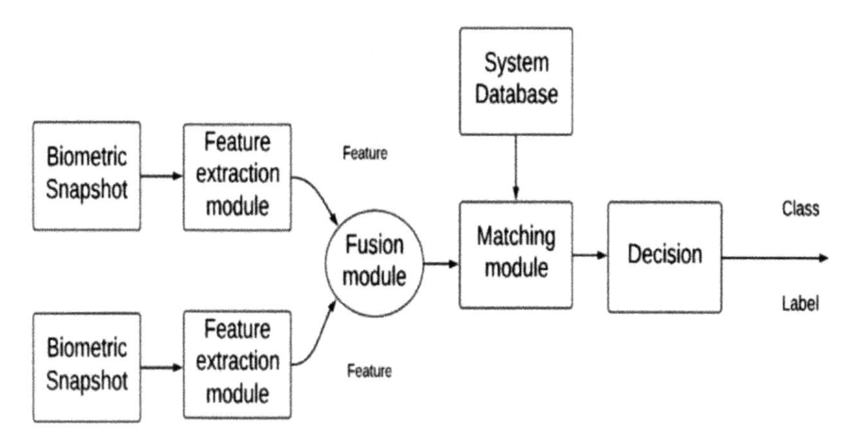

FIGURE 7.5 Biometric fusion.

Features contain more extravagant data of biometric qualities when compared with matching score or choice of matcher and in this manner fusion at the component level gives better acknowledgment comes about. It has additionally been observed that feature level fusion gives more exactness when the features of different biometric modalities are good with each other.

7.9.7 MATCHING SCORE LEVEL FUSION

In this fusion level, the feature vectors are handled independently rather than processing it independently. At that point, an individual matching score is found and in view of the exactness of each biometric channel, then at that point intertwine the coordinating level to and a composite matching score which will be utilized for classification. We can utilize different procedures, for example, logistic regression, most high rated rank, Bayes rules, mean fusion, and so forth, to consolidate match scores. Moreover, another vital part of this fusion is the normalization of scores gained from different modalities. We can utilize strategies, for example, Min-max, z-score, piecewise linear, and so forth, to accomplish normalization of the match scores. Matching score level fusion has lesser complex nature than the other fusion levels and henceforth it is broadly utilized (Fig. 7.6).

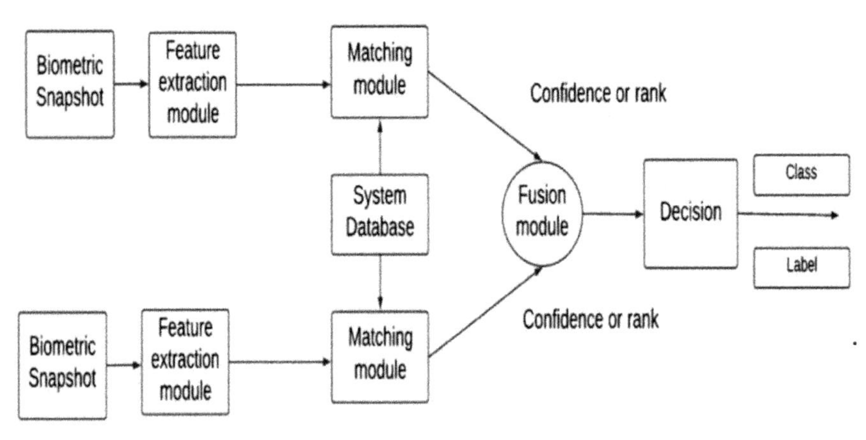

FIGURE 7.6 Biometric fusion score levels.

7.9.8 DECISION-LEVEL FUSION

In decision level fusion, each biometric characteristic is first preclassified independently. The individual biometric attribute is first caught and after that features are separated from the caught characteristic. The attributes are classified as either acknowledge or dismiss in light of these separated features. The final classification is acquired by consolidating the yields of different modalities (Fig. 7.7).

7.9.9 ADVANTAGES OF MULTIMODAL BIOMETRIC SYSTEMS

The accuracy of a multimodal biometric system is estimated by the errors in image acquisition and matching of the biometric characteristics. Image acquisition includes failure-to-acquire (FTA) rate and FTE rate. Matching errors comprise false non-match rates (FNMR) in which a legitimate subject is rejected and a false match rate (FMR) where an intruder is granted access. Multimodal systems have around zero FTA, FTE, FNMR, and FMR rates.

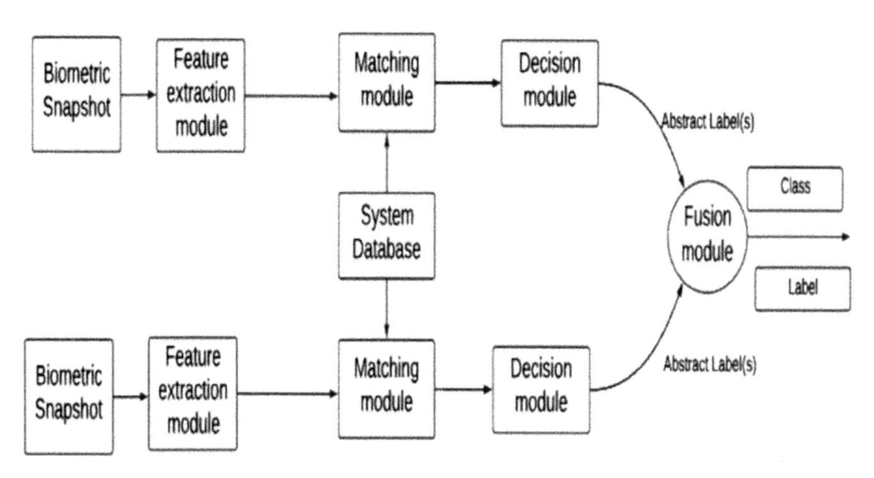

FIGURE 7.7 Biometric fusion module.

Multimodal biometrics can diminish information distortion. In situations where the quality of a biometric test is acceptable, the other biometric attribute can be utilized. For instance, if a fingerprint scanner rejects the fingerprint image because of low quality and uses another biometric methodology, for example, facial rejection will bring down the false rejection rates (FRR). Multimodal biometric systems are extremely difficult to spoof when compared with unimodal systems. Regardless of whether one biometric methodology could be spoofed, the individual can, in any case, be verified utilizing the other biometric identifier.

The following are the advantages that are offered by the multimodal biometric system more than that of the unimodal biometric system:

1. As we know unimodal biometric system acquires only one trait during recognition, whereas the multimodal biometric system uses more than one type of information and also provides an improvement in the matching accuracy. Multimodal biometric systems are capable of solving a greater number of nonuniversality issues that a wide population of users require. In the multimodal biometric system if the user doesn't have a particular trait at any time he/she can switch to any other biometric trait to enter into the system, whereas in the unimodal biometric system we have to rely upon only one of the trait that was registered. Only part of

the information enrolled by a user is requested for verification and a certain degree of flexibility is obtained by enrolling the user by acquiring his multiple traits.

2. It becomes very challenging to bluff the legitimate user who has been enrolled in the multimodal biometric system as these are less delicate to imposter attacks.
3. When information acquired from the single biometric trait is falsified by noise then another trait of the same user could be used to perform the verifications as multimodal systems are insensitive to noise on the sensors.

When a particular biometric trait is not adequate in continuously monitoring or tracking the person in specific conditions these systems can prove helpful. For example, tracking a person using face and gait simultaneously.

7.9.10 NEED OF MULTIMODAL BIOMETRICS

In today's world, most biometric systems are established such that applications support single proof of supply of data for authentication (e.g., face, fingerprint, voice, etc.) which are unimodal. These systems respond area unit unsafe attribute to the incidence of different kinds of issues, such as clanging data, intra class variations, interclass similarities, non-universality and spoofing as a result it ends up having a high false acceptance rate (FAR) and FRR, restrictive discrimination competence and lack of endurance. In order to provide identity, there are some limitations that are advertised by the unimodal biometric system must overcome through combining multiple sources of data together. A pair of two or numerous styles of a biometric system is referred to as multimodal biometric system in which integration of area modal is allowed.

Data and the efficiency of generally higher intellectual {process} process may be heightened by blending multiple procedures, for example, the efficiency of police investigation events from a team sports video has solely become feasible by blending audio-visual options along with that of matter info. Multimodal fusion is essential however with a precise value and problem within the analysis method.

7.9.11 IRIS AND FINGERPRINT BASED MULTIMODAL BIOMETRICS SYSTEM

During the analysis of multimodal biometric system iris and fingerprint, traits are combined together for the analysis.

Iris and fingerprint traits are here combined together for the analysis of the multimodal biometric system. This chapter describes the architecture which uses wavelet and texture-based feature extraction method.

The multimodal biometric system uses two biometric traits such as iris and fingerprints. Therefore, for both of these traits the process is as follows:

Capturing the biometric trait sample where numerous samples have been collected for both the preprocessing phase where each sample has been normalized and converted into the grayscale as required and then feature extraction is applied.

7.9.12 FEATURE EXTRACTION PHASE

1. Iris feature extraction: in this phase sample features are collected and then these are converted into grayscale after this localization and texture features has been extracted. These features are saved and then extrema centroid and area features are extracted.
2. Fingerprint extraction: in this phase sample features are collected and then these are converted into grayscale after this localization and texture features has been extracted. These features are saved and then extrema centroid, perimeter, convex hull, maxima, and minima features are extracted.

7.10 STAGES IN BIOMETRIC RECOGNITION SYSTEM

7.10.1 PREPROCESSING STAGE

As depicted in Figure 7.8, the following are the stages involved in the process of a biometric system:

- pre-processing stage,
- feature extraction stage,
- matching, and
- decision stage.

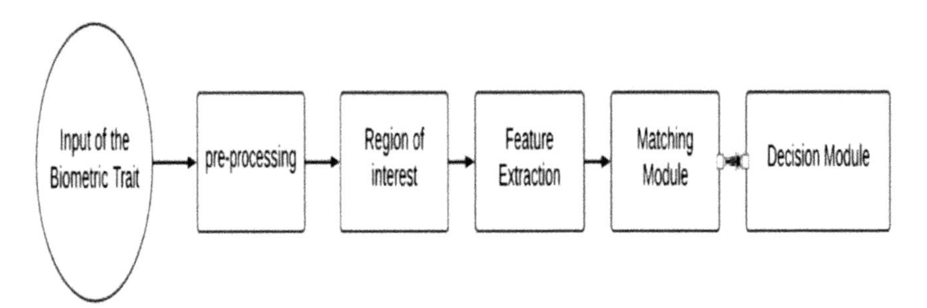

FIGURE 7.8 Biometric stages.

Preprocessing stage consists of the acquisition of data from the sensor with the help of computer peripherals, such that there is a significantly large improvement in the quality of the biometric data recognition.

Rehman and Saba[5] together worked on neural network for document image processing. The leverage on their work was due to the fact that neural networks are widely known and are used widely in biometrics to implement and perform a crucial function in document layout analysis and classification.

These functionalities are used to intensify the actual image for future analysis by breaking these down into minute tasks, such as removal of lines, skew estimation, correction, and baseline detection, etc.

Goh et al.[6] proposed a wavelet local binary fusion which is shown as illuminated facial image preprocessing for face recognition. Han et al.[7] performed a study on illumination preprocessing in face recognition and clubbed them into three main categories based on different principles, such as

- gray-level transformation,
- gradient or edge extraction, and
- reflectance field estimation.

The experiments resulted that there is some improvement in the performance due to some of the holistic illumination preprocessing techniques.

Han et al.[8] used five steps to achieve a high-resolution fingerprint image that includes normalization, orientation image estimation, frequency image estimation, region mask generation, and filtering using Gabor filter.

Han showed that with the help of these steps we can possibly detect all the minutiae present in the fingerprint images and these are then used for further identification.

Utilizing these steps, results demonstrated that it was conceivable to distinguish all the minutiae exhibit in the unique finger impression image, which is utilized for identification.[9] He described four steps for the preprocessing stage in the construction of a face recognition system. The steps included gamma correction, difference of Gaussian filtering, masking, and equalization of variation. To expel noise in a hand composed signature verification system utilizing a median filter, image binarization utilizing Otsu binarization algorithm and cropping signature the accompanying prepreparing steps were utilized normalization, resampling, smoothing, and image enhancement.[10] To enhance the performance of an automated speech recognition system, the input speech was provided through the linear predictive analysis filter. Tests utilizing standard feature vectors demonstrated that the linear predictive filter enhances robustness in speech recognition system.[11] For an iris biometric recognition system, Lili et al.[12] proposed a preprocessing procedure by presenting an iris localization algorithm and likewise adopted the edge points detecting and curve fitting. Tests on 2400 iris images from the Institute of Automation, Chinese Academy database demonstrated that every one of the methodologies utilized was substantial. For a secured personal identification system in light of human retina, Fatima et al.[13] proposed a preprocessing strategy that excerpts the vascular pattern from input retina image utilizing wavelets and multilayered threshold technique. In biometric online signature verification, Lopez-Garcia et al.[14] preprocessed the data to decrease noise and normalize the signature stroke by following these four stages; filtering, equally-spacing, location, and time normalization, and size normalization. For enhanced sensor interoperability in online biometric signature verification, Tolosana et al.[15] proposed a preprocessing stage where information gained from various gadgets are handled with a specific end goal to normalize the signs in similar ranges.[38–52]

7.10.2 REGION OF INTEREST

The Region of Interest (ROI) in biometric systems is a procedure implanted inside or a procedure done before moving into the feature extraction stage. The ROI is the process of featuring key and interesting features in a biometric trait as a smaller region that will additionally be utilized as matching attributes in a biometric system.[16] In palmprint recognition, researches uncover that finite sizes are utilized to get the

ROI while different investigations expressed that a region estimate is extricated, which is simply the measure of the palm. Be that as it may, the different ROI strategies are reliable on the database of decision. Thus, the database utilized for the diverse modalities of biometric systems is additionally a main consideration that will decide the ROI extraction strategy to be utilized.[17] The ROI strategies can be characterized into four broad categories to be specific:

1. Bottom-up feature-based approaches: This approach expects to find features where there are fluctuating conditions in lighting, posture, and so forth, and make utilization of them in the detection strategies. Such an approach incorporates the scale-invariant feature transform (SIFT) which is significantly utilized for the face localization. SIFT works by assembling a scale-invariant region identifier and a descriptor in light of the gradient distribution in the detected regions. Likewise, although the SIFT method is the shape context method that has a similar thought, it depends on edges extricated by the canny detector where the location is quantized into nine bins of a log-polar coordinate system. Moreover, another comparable technique is the histograms of oriented gradients that utilizes overlapping neighborhood contrast normalization for enhancing its exactness and it is computed on a dense grid of consistently spread cells, demonstrating that utilizing a linear support vector machine it can help in classification of human beings.[18]
2. Top-down knowledge-based approach: This approach can control the false positive circumstances of situations as it is significantly concerned about the object of interest. Viola and Jones[19] proposed a ROI approach that considers existing data on a person's movement and an appearance by choosing of capabilities, size of the preparation information, and scales utilized for discovery. Thus, it does not support an alternate strategy of tracking and arrangement.
3. Template matching approach: This approach utilizes object parts to represent the global image of the object where because of this association; there is a plausibility for calculation which is additionally utilized for detection. Gagan et al.[20] proposed a human detection idea that differentiates foreground regions and extracts the boundary. The algorithm additionally scans for people within an image by matching edge features of human outline stored inside the database.

4. Appearance-based approach: This approach includes learning models from a preparatory set of images that are additionally utilized for detection. The approach relies on machine learning and statistical analysis procedure to locate unique kind of attributes of images containing objects. These learned characteristics which are in the form of discriminant functions are additionally utilized for detection.[18] Diaz et al.[21] proposed a novel technique in view of orientation entropy for incomplete fingerprint keeping in mind the end goal to separate the ROI and utilizing Poincare index to detect the reference point. Trials done on the FVC 2004 database demonstrated that the approach found the position of reference of all type of fingerprints. Saliha et al.[22] proposed a ROI extraction method for the palm that contained its uniquely distinguished features. Key point localization was utilized to aim for finding crucial points that keeping in mind the end goal to adjust the hand image. The technique was additionally in view of incorporating the x-axis projection and the projection of the upper and lower edge, by permitting these lines the extraction of the horizontal limits reaches of the hand contour. A 3D face recognition was produced by assembling the whole face with the regions of interest, that is, mouth, nose, left eye, and the right eye, where Gabor filter was utilized to improve the discriminant data stages and the principal component analysis was incorporated into the information to acquire a decreased basis projection and discriminant.[32]

7.11 FEATURE EXTRACTION STAGE

Feature extraction is likewise a vital stage in the identification procedure of biometric systems. It includes the rendering of the chunk of resources that describes an extensive large set of information. Feature extraction is primarily used to limit the initial dataset by getting a few properties that can be utilized to characterize and get patterns that are available in the input images.[24] For an offline handwritten signature system, two feature extraction approaches were proposed called the static and pseudodynamic methodologies. The primary approach included estimating the geometrical features whereas in the second approach, the estimation of dynamic information of the image was attempted.[17] For extraction of features in an ear biometric recognition system, Khobragade et al.[25] proposed the sparse portrayal of ear biometrics and Ghoualmi[26] proposed an invariant technique for evolution and internal ear curves. A technique for feature extraction called pulse active ratio, which is created in light of the rule of pulse width modulation, was actualized on the electrocardiogram signal for biometric authentication.[27]

Gabor filter, discrete wavelet transform, discrete cosine transform, and fast Fourier transform were strategies utilized for feature extraction on a fingerprint recognition system. Experimental observations were carried out on the FVC 2002 database demonstrated that DCT and FFT performed better contrasted with DWT and Gabor filter.[28] Three procedures to be specific wavelet transform based feature extraction, spatial differentiation, and twin pose testing scheme were utilized as feature extraction methods to enhance the performance of a face recognition system.[29] To find the optic circle of the retina, Radha et al.[30] utilized the region threshold and the presence of contours was seen by methods for the Hough transform. For speech recognition system, wavelet-based feature extraction strategy for speech data was presented. Test demonstrated that the proposed feature extraction strategy gave a high recognition rate.[31] To enhance the performance of voice recognition system, Chauhan[32] proposed the Mel frequency cepstral coefficients as a feature extraction strategy that incorporates wiener filter that is used to deal with the noise in speech. To get iris feature from the iris image, Ukpai et al.[33] proposed an iris feature extraction strategy that depends on the principle texture and dual-tree complex wavelet transform.

Likewise, Lahroodi et al.[34] displayed the utilization of Laplacian-of-Gaussian filters to acquire the isotropic band-pass deterioration of the normalized iris image for feature extraction. A median filter was utilized as a smoothing method to evacuate noise in the border of the hand region in a hand geometry recognition system[35] and from that point, the Competitive Hand valley detection was utilized to find the valley points. Utilizing high-resolution approach for palmprint biometric system, features, for example, orientation, thickness, principal lines, and minutiae are taken as feature extraction.[36] The matching module stage is the place the feature values are compared with the features present in the template in this manner producing a matching score. A case in this module is the point at which the facial features separated are contrasted and the features shown in the template database, which is additionally computed and treated as a matching score. Likewise, the decision-making module is that phase where the identity of the client is either acknowledged or dismissed with respect to the matching score produced in the matching module.[37]

KEYWORDS

- **biometric system**
- **unimodal biometric**
- **multimodal biometric**
- **fingerprint**
- **minutiae**
- **spoof assaults**

REFERENCES

1. Os, E. D.; Jongebloed, H.; Stijsiger, A.; Boves, L. In *Speaker Verification as a User-friendly Access for the Visually Impaired*, Proceedings of the Sixth European Conference on Speech Technology, Budapest, Hungary, 1999, pp 1263–1266.
2. Eriksson, A.; Wretling, P. In *How flexible is the Human Voice? A Case Study of Mimicry*, Proceedings of the European Conference on Speech Technology, Rhodes, 1997, pp 1043–1046.
3. Harrison, W. R. Suspect Documents, Their Scientific Examination. Chicago, IL: Nelson-Hall, 1981.
4. Matsumoto, T.; Matsumoto, H.; Yamada, K.; Hoshino, S. Impact of Artificial Gummy Fingers on Fingerprint Systems. *Proc SPIE* **2002,** *4677*, 275–289.
5. Rehman, A.; Saba, T. Neural Networks for Document Image Preprocessing: State of the Art. *Artificial Intell. Rev.* **2014,** *42*, 253–273.
6. Goh, Y. Z.; Teoh, A. B. J.; Goh, M. K. O. Wavelet Local Binary Patterns Fusion as Illuminated Facial Image Preprocessing for Face Verification. *Expert Syst. Appl.* **2011,** *38*, 3959–3972.
7. Han, H.; Shan, S.; Chen, X.; Gao, W. A Comparative Study on Illumination Preprocessing in Face Recognition. *Pattern Recog.* **2013,** *46*, 1691–1699.
8. Nava, S. G.; Silva, A. J. R.; Montiel, N. H.; Funes, F. J. G.; González, M. D. In *Hybrid Methodology Focused on the Model of Binary Patterns and the Theory of Fuzzy Logic for Facial Biometric Verification and Identification*, 2015 7th International Joint Conference on Computational Intelligence (IJCCI), 2015, pp 180–187.
9. Jahanbin, S.; Choi, H.; Bovik, A. C. Passive Multimodal 2-D+ 3-D Face Recognition Using Gabor Features and Landmark Distances. *IEEE Trans. Inform. Forensics Security* **2011,** *6*, 1287–1304.
10. Mohammed, R. A.; Nabi, R. M.; Sardasht, M.; Mahmood, R.; R. M. Nabi *State-of-the-Art in Handwritten Signature Verification System*, International Conference on Computational Science and Computational Intelligence (CSCI), 2015, pp 519–525.
11. Koniaris, C.; Chatterjee, S. *A Sparsity Based Preprocessing for Noise Robust Speech Recognition*, Spoken Language Technology Workshop (SLT), 2014, 513–518.
12. Lili, P.; Mei, X. *The Algorithm of Iris Image Preprocessing*, Fourth IEEE Workshop on Automatic Identification Advanced Technologies (AutoID'05), 2005, 134–138.

13. Fatima, J.; Syed, A. M.; Akram, M. U. *A Secure Personal Identification System Based on Human Retina*, IEEE Symposium on Industrial Electronics and Applications (ISIEA), 2013, 90–95.
14. López-García, M.; Ramos-Lara, R.; Miguel-Hurtado, O.; Cantó-Navarro, E. Embedded System for Biometric Online Signature Verification, *IEEE Trans. Ind. Inform.* **2014,** *10*, 491–501.
15. Tolosana, R.; Vera-Rodriguez, R.; Ortega-Garcia, J.; Fierrez, J. Preprocessing and Feature Selection for Improved Sensor Interoperability in Online Biometric Signature Verification. *IEEE Access* **2015,** *3*, 478–489.
16. Barbosa, F. G.; Silva, W. L. S. *Automatic Voice Recognition System Based on Multiple Support Vector Machines and Mel-frequency Cepstral Coefficients*, 2015 11th International Conference on Natural Computation (ICNC) 2015, pp 665–670.
17. Poon, C.; Wong, D. C.; Shen, H. C. In *A New Method in Locating and Segmenting Palmprint into Region-of-Interest*, Proceedings of the 17th International Conference on Pattern Recognition, 2004, pp 533–536.
18. Shekhar, S.; Kumar, B. S.; Ramesh, S. *Robust Approach for Palm (Roi) Extraction in Palmprint Recognition System*, IEEE International Conference on Engineering Education: Innovative Practices and Future Trends (AICERA), pp 1–6.
19. Jones, M. J.; Viola, P. *Method and System for Object Detection in Digital Images*, ed: Google Patents, 2006.
20. Gagan, R.; Lalitha, S.; *Elliptical Sector Based DCT Feature Extraction for Iris Recognition*, IEEE International Conference on Electrical, Computer and Communication Technologies (ICECCT), 2015, pp 1–5.
21. Diaz, M. R.; Travieso, C. M.; Alonso, J. B.; Ferrer, M. A. *Biometric System Based in the Feature of Hand Palm*, International Carnahan Conference on Security Technology, 2004, pp 136–139.
22. Saliha, A.; Karima, B.; Mouloud, K.; Nabil, D. H. Ahmed, B. *Extraction Method of Region of Interest from Hand Palm: Application with Contactless and Touchable Devices*, International Conference on Information Assurance and Security (IAS), 2014, pp 77–82.
23. Belahcene, M.; Chouchane, A.; Ouamane, H. *3D Face Recognition in Presence of Expressions by Fusion Regions of Interest*, 22nd Signal Processing and Communications Applications Conference (SIU), 2014, pp 2269–2274.
24. Fasel, B.; Luettin, J. *Automatic Facial Expression Analysis: A Survey. Pattern Recog.* **2003,** 36, 259–275.
25. Khobragade, S.; Mor, D. D.; Chhabra, A. *A Method of Ear Feature Extraction for Ear Biometrics Using MATLAB*, Annual IEEE India Conference (INDICON), 2015, pp 1–5.
26. Ghoualmi, L.; Draa, A.; Chikhi, S. *An Efficient Feature Selection Scheme Based on Genetic Algorithm for Ear Biometrics Authentication*, 12th International Symposium on Programming and Systems (ISPS), 2015, pp 1–5.
27. Soni, K.; Gupta, S. K.; Kumar, U. Agrwal, S. L. *A New Gabor Wavelet Transform Feature Extraction Technique for Ear Biometric Recognition*, 6th IEEE International Conference (PIICON) on Power India, pp 1–3.
28. Tewari, K.; Kalakoti, R. L. *Fingerprint Recognition and feature extraction using transform domain techniques*, International Conference on Advances in Communication and Computing Technologies (ICACACT), 2014, 1–5.

29. Kumar, T. S.; Kanhangad, V. *Face Recognition Using Two- Dimensional Tunable-Q Wavelet Transform,* International Conference on Digital Image Computing: Techniques and Applications (DICTA), 2015.
30. Radha, R.; Lakshman, B. *Identification of Retinal Image Features Using Bitplane Separation and Mathematical Morphology*, World Congress on Computing and Communication Technologies (WCCCT), 2014, pp 120–123.
31. Gunal, S.; Edizkan, R. *Use of Novel Feature Extraction Technique with Subspace Classifiers for Speech Recognition,* IEEE International Conference on Pervasive Services, 2007, 80–83.
32. Chauhan, P. M.; Desai, N. P. *Mel Frequency Cepstral Coefficients (MFCC) Based Speaker Identification in Noisy Environment Using Wiener Filter*, International Conference on Green Computing Communication and Electrical Engineering (ICGCCEE), 2014, pp 1–5.
33. Ukpai, C. O.; Dlay, S.; Woo, W. *Iris Feature Extraction Using Principally Rotated Complex Wavelet Filters (PR-CWF)*, International Conference on Computer Vision and Image Analysis Applications (ICCVIA), 2015, pp 1–6.
34. Lahroodi, M.; Konukseven, E. I. *New Parameter Estimation Method of Linear Adaptive Filter for Modeling Kinematic Motion Primitive of Effective DOF of Hand*, IEEE Conference on Control Applications (CCA), 2015, pp 157–162.
35. Nguyen, T. -N.; Vo, D. -H.; Huynh, H. -H.; Meunier, J. *Geometry-based Static Hand Gesture Recognition Using Support Vector Machine*, 13th International Conference Control Automation Robotics & Vision (ICARCV), pp 769–774.
36. Sanyal, N.; Chatterjee, A.; Munshi, S. *A Novel Palmprint Authetication System by XWT Based Feature Extraction and BFOA Based Feature Selection and Optimization*, 2nd International Conference on Recent Trends in Information Systems (ReTIS), pp 455–460.
37. Ross, A.; Jain, A. Information Fusion in Biometrics. *Pattern Recog. Let.* **2003,** *24*, 2115–2125.
38. Chowdhary, C. L.; Acharjya, D. P. Singular Value Decomposition–Principal Component Analysis-Based Object Recognition Approach. In *Bio-Inspired Computing for Image and Video Processing;* Taylor & Francis Group: Oxford shire, United Kingdom, 2018; p 323.
39. Chowdhary, C. L. Application of Object Recognition with Shape-Index Identification and 2D Scale Invariant Feature Transform for Key-Point Detection. *Feature Dimension Reduction for Content-Based Image Identification;* IGI Global: USA, 2018; pp 218–231.
40. Chowdhary, C. L.; Muatjitjeja, K.; Jat, D. S. *Three-dimensional Object Recognition Based Intelligence System for Identification*, Conference on Emerging Trends in Networks and Computer Communications (ETNCC), 2015.
41. Chowdhary, C. L.; Ranjan, A.; Jat, D. S. *Categorical Database Information-Theoretic Approach of Outlier Detection Model*, Annals. Computer Science Series. 14th Tome 2nd Fasc. 2016, pp 29–36.
42. Chowdhary, C. L. Linear Feature Extraction Techniques for Object Recognition: Study of PCA and ICA. *J. Serbian Soc. Comput. Mech.* **2011,** *5* (1), 19–26.

43. Chowdhary, C. L.; Acharjya, D. P. Breast Cancer Detection Using Hybrid Computational Intelligence 43. Techniques. *Handbook of Research on Emerging Perspectives on Healthcare Information Systems and Informatics;* 2018; pp 251–280.
44. Chowdhary, C. L.; Acharjya, D. P. Segmentation of Mammograms Using a Novel Intuitionistic Possibilistic Fuzzy C-Mean Clustering Algorithm. *Nature Inspired Computing;* Springer: Singapore, 2018; pp 75–82
45. Chowdhary, C. L.; Acharjya, D. P. *Clustering Algorithm in Possibilistic Exponential Fuzzy c-mean Segmenting Medical Images. J. Biomim. Biomat. and Biomed. Eng.* 2017, *30*, 12–23.
46. Chowdhary, C. L.; Acharjya, D. P. A Hybrid Scheme for Breast Cancer Detection Using Intuitionistic Fuzzy Rough Set Technique. *Biometrics: Concepts, Methodologies, Tools, and Applications;* 2016; pp 1195–1219.
47. Das, T. K. & Chowdhary, C. L. Implementation of Morphological Image Processing Algorithm Using Mammograms. *J. Chem. Pharm. Sci.* **2016,** *10* (1), 439–441.
48. Chowdhary, C. L. A Review of Feature Extraction Application Areas in Medical Imaging. *Int. J. Pharm. Technol.* **2016,** *8* (3), 4501–4509.
49. 49. Chowdhary, C. L.; Acharjya, D. P. In *Breast Cancer Detection using Intuitionistic Fuzzy Histogram Hyperbolization and Possibilitic Fuzzy c-mean Clustering Algorithms with Texture Feature Based Classification on Mammography Images*, Proceedings of the International Conference on Advances in Information Communication Technology & Computing, 2016, p 21.
50. Chowdhary, C. L.; Sai, G. V. K.; Acharjya, D. P. Decreasing False Assumption for Improved Breast Cancer Detection. *J. Sci. Arts* **2016,** *35* (2), 157–176.
51. Chowdhary, C. L.; Sai, G. V. K.; Acharjya, D. P. Decrease in False Assumption for Detection Using Digital Mammography. *Computational Intelligence in Data Mining;* Springer, New Delhi, 2015; Vol. 2, pp 325–333.
52. Chowdhary, C. L. *Appearance-based 3-D Object Recognition and Pose Estimation: Using PCA, ICA and SVD-PCA Techniques*, LAP Lambert Acad: Germany, 2011.

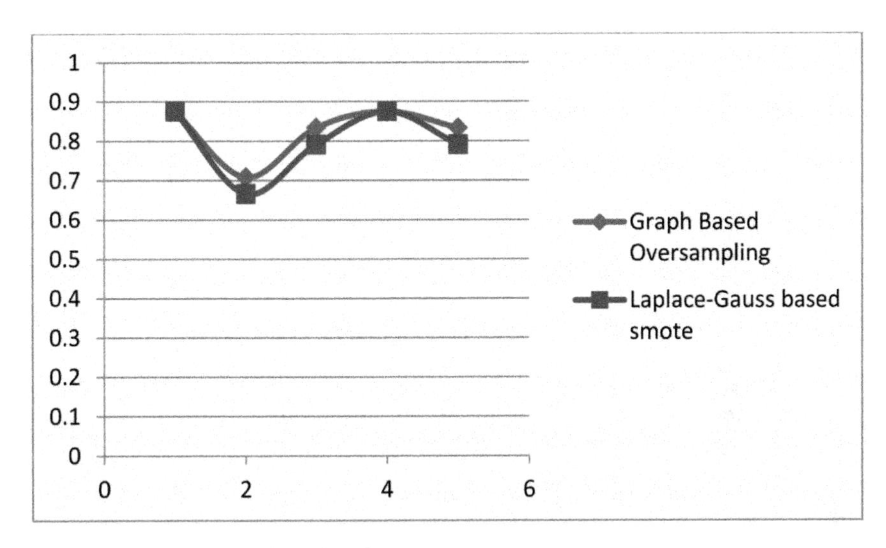

FIGURE 8.1 Mean ranking of performance measures.

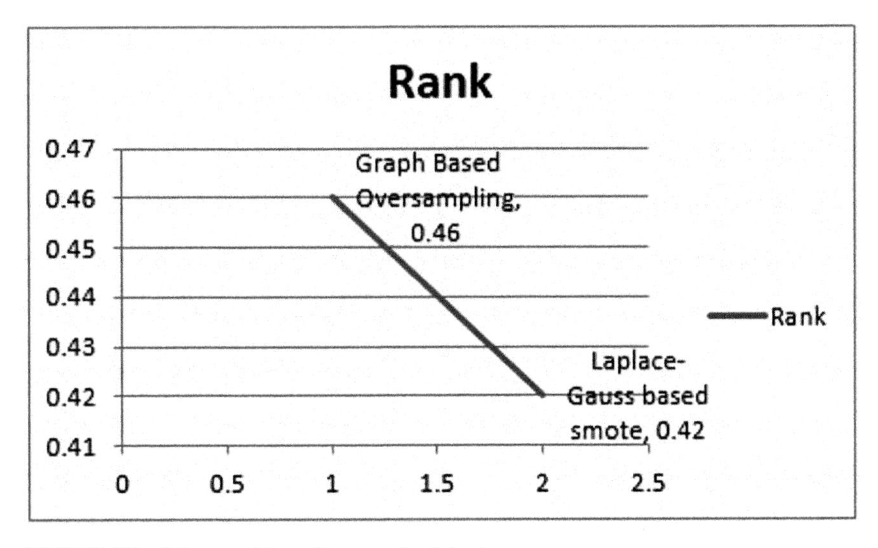

FIGURE 8.2 Mean ranking of two methodologies.

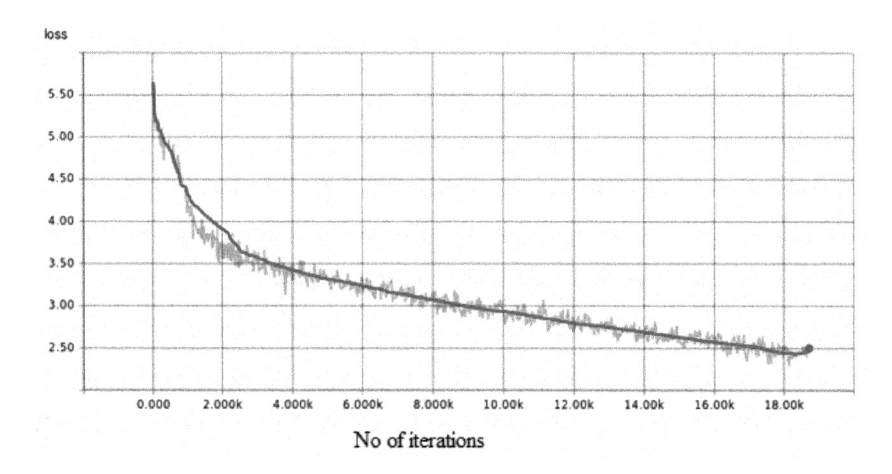

FIGURE 12.3 Loss minimizer.

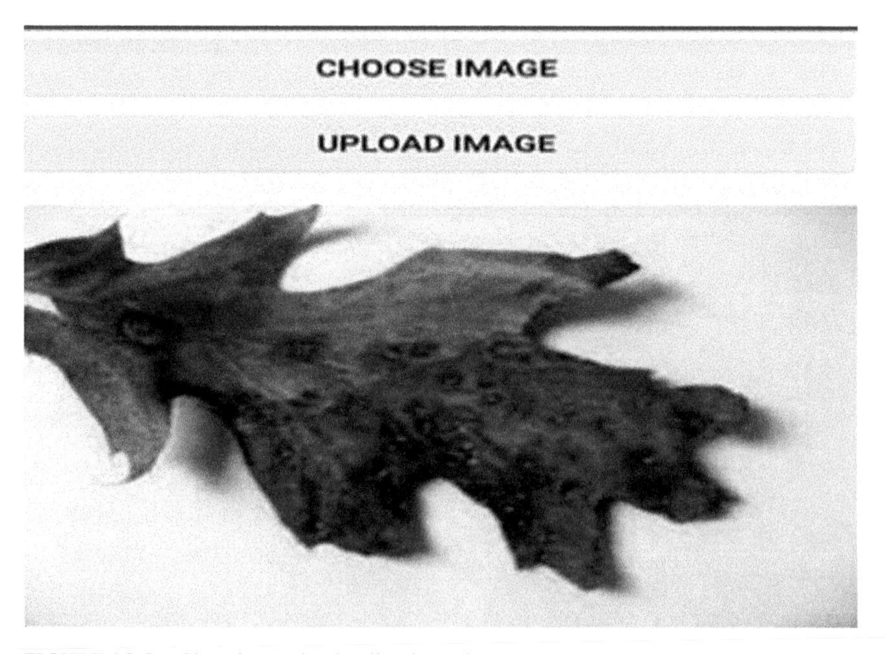

FIGURE 13.3 Choosing and uploading input image.

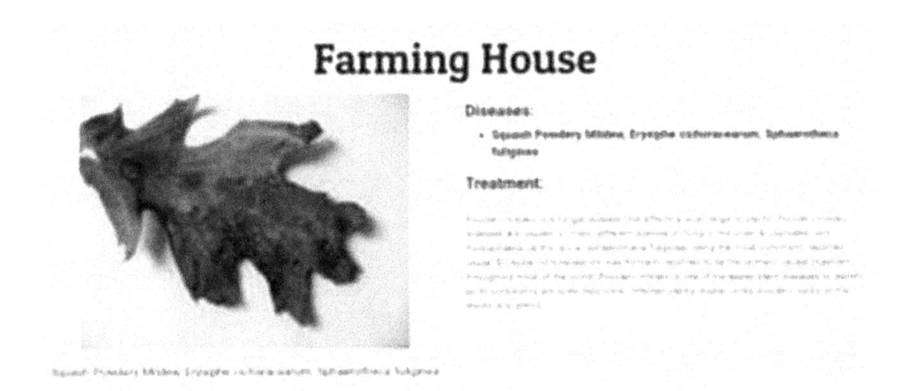

FIGURE 13.4 Identifying disease and sharing details.

FIGURE 14.3 Preparing dataset.

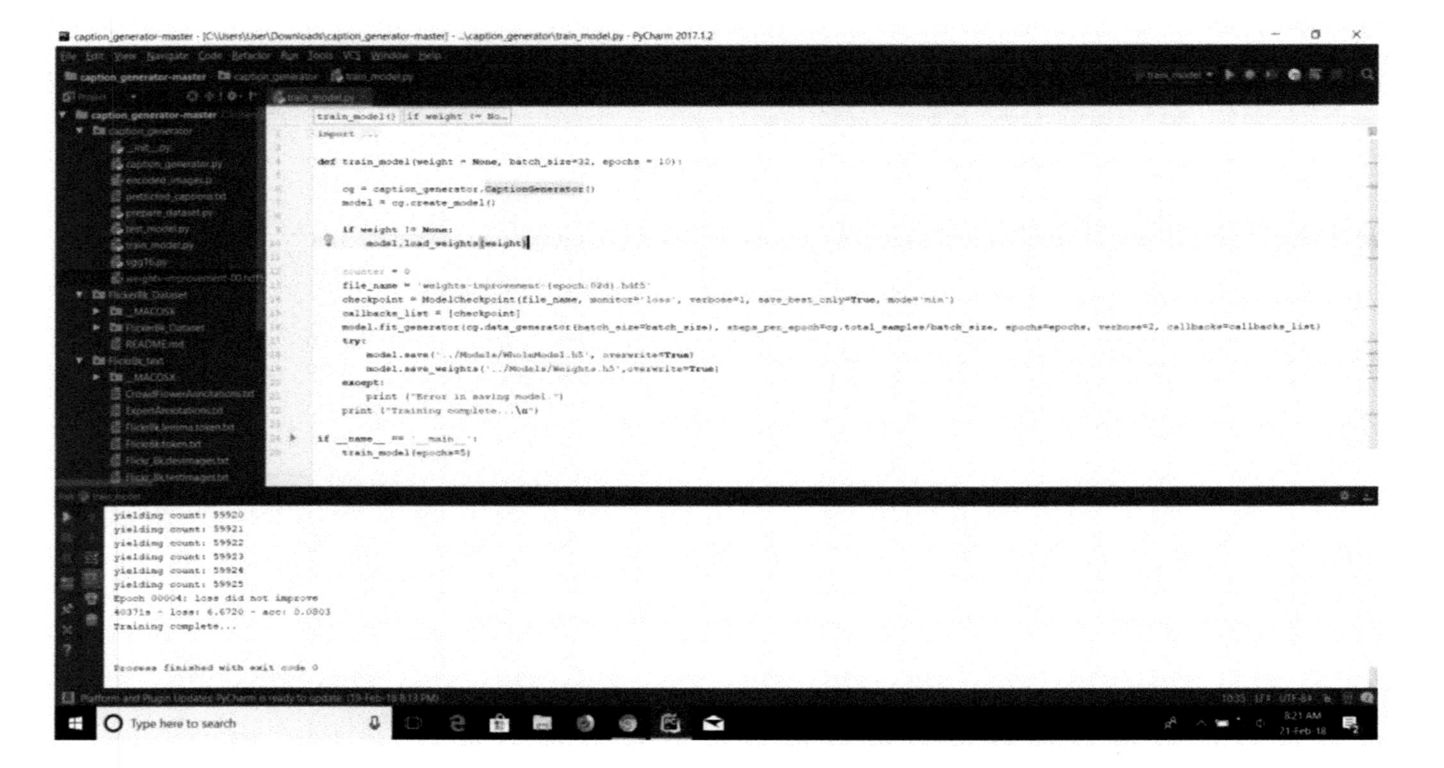

FIGURE 14.4 Training.

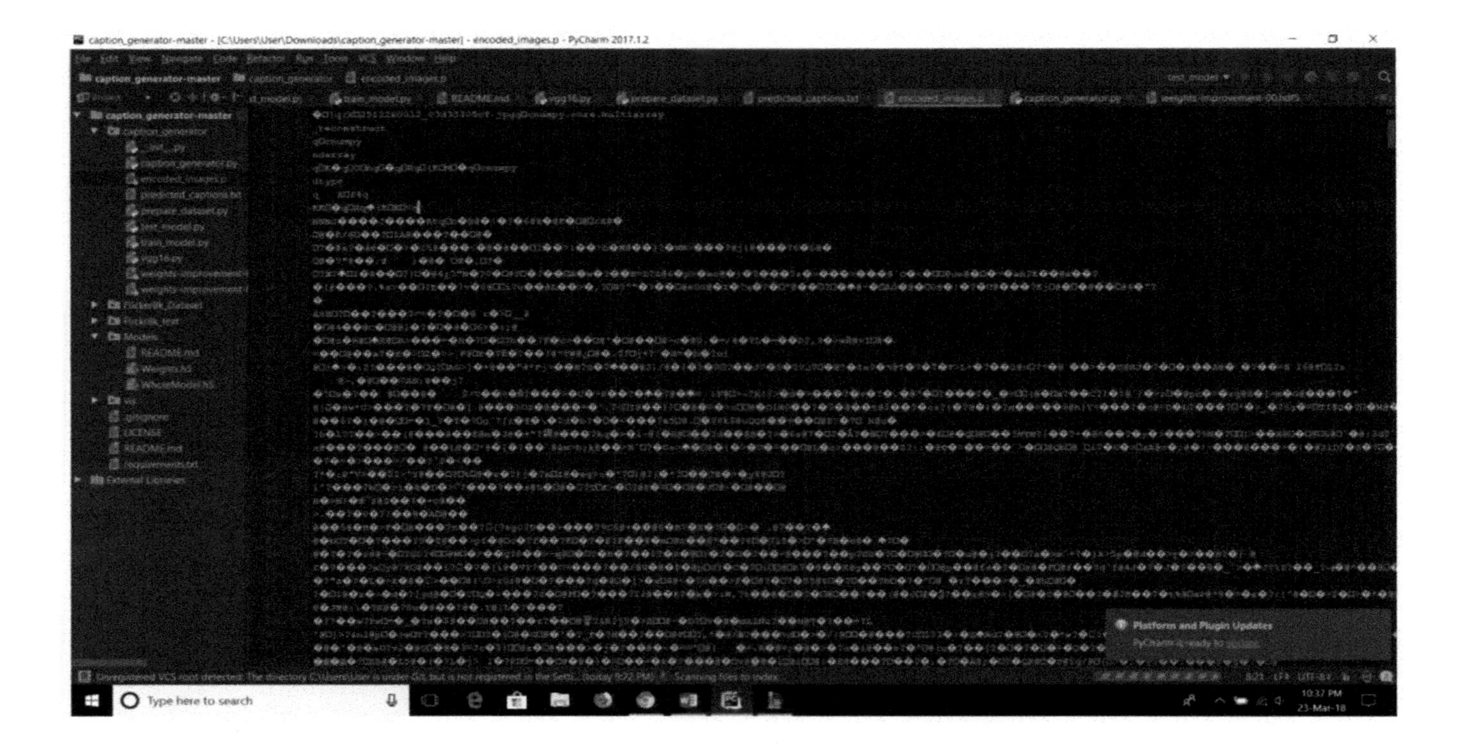

FIGURE 14.5 Testing.

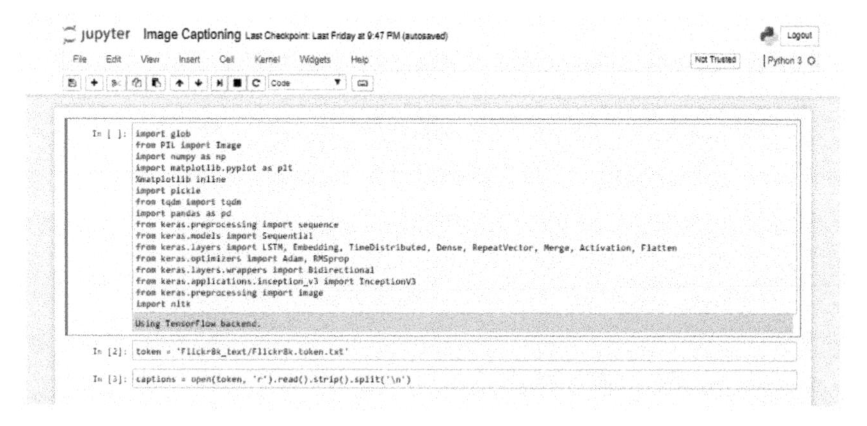

FIGURE 14.6 Preparing dataset.

Creating a dictionary containing all the captions of the images

```
In [4]: d = {}
        for i, row in enumerate(captions):
            row = row.split('\t')
            row[0] = row[0][:len(row[0])-2]
            if row[0] in d:
                d[row[0]].append(row[1])
            else:
                d[row[0]] = [row[1]]

In [5]: d['1000268201_693b08cb0e.jpg']

Out[5]: ['A child in a pink dress is climbing up a set of stairs in an entry way .',
         'A girl going into a wooden building .',
         'A little girl climbing into a wooden playhouse .',
         'A little girl climbing the stairs to her playhouse .',
         'A little girl in a pink dress going into a wooden cabin .']

In [6]: images = 'Flickr8k_Dataset/Flicker8k_Dataset/'

In [7]: # Contains all the images
        img = glob.glob(images+'*.jpg')

In [8]: img[:5]

Out[8]: ['Flickr8k_Dataset/Flicker8k_Dataset/17273391_55cfc7d3d4.jpg',
         'Flickr8k_Dataset/Flicker8k_Dataset/2890075175_4bd32b201a.jpg',
         'Flickr8k_Dataset/Flicker8k_Dataset/3356642567_f1d92cb81b.jpg',
         'Flickr8k_Dataset/Flicker8k_Dataset/186890605_ddff5b694e.jpg',
         'Flickr8k_Dataset/Flicker8k_Dataset/2773682293_3b712e47ff.jpg']

In [9]: train_images_file = 'Flickr8k_text/Flickr_8k.trainImages.txt'

In [10]: train_images = set(open(train_images_file, 'r').read().strip().split('\n'))

In [11]: def split_data(l):
             temp = []
             for i in img:
                 if i[len(images):] in l:
                     temp.append(i)
             return temp

In [12]: # Getting the training images from all the images
         train_img = split_data(train_images)
         len(train_img)
```

FIGURE 14.7 Creating a dictionary for captions.

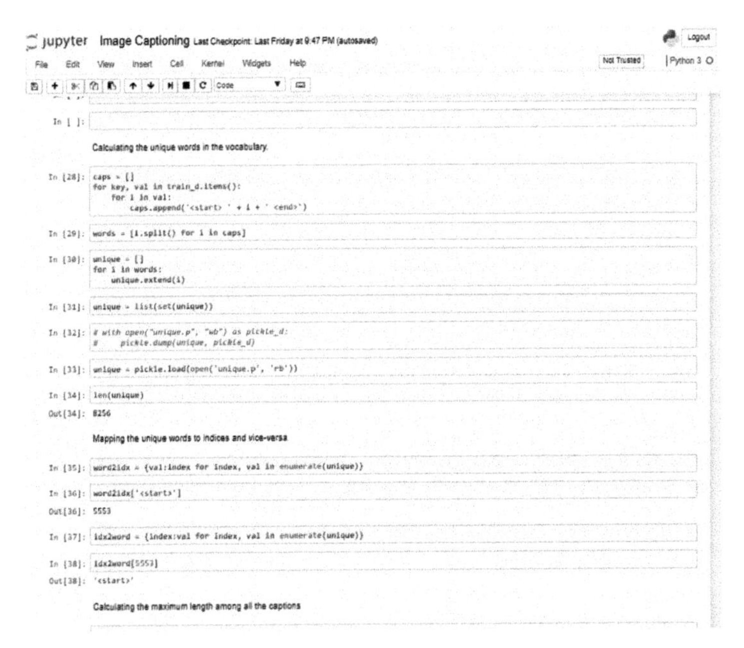

FIGURE 14.8 Mapping of unique words.

upyter Image Captioning Last Checkpoint: Last Friday at 9:47 PM (autosaved)

Edit View Insert Cell Kernel Widgets Help Not Trusted Python

Code

```
In [47]: def data_generator(batch_size = 32):
             partial_caps = []
             next_words = []
             images = []

             df = pd.read_csv('flickr8k_training_dataset.txt', delimiter='\t')
             df = df.sample(frac=1)
             iter = df.iterrows()
             c = []
             imgs = []
             for i in range(df.shape[0]):
                 x = next(iter)
                 c.append(x[1][1])
                 imgs.append(x[1][0])


             count = 0
             while True:
                 for j, text in enumerate(c):
                     current_image = encoding_train[imgs[j]]
                     for i in range(len(text.split())-1):
                         count+=1

                         partial = [word2idx[txt] for txt in text.split()[:i+1]]
                         partial_caps.append(partial)

                         # Initializing with zeros to create a one-hot encoding matrix
                         # This is what we have to predict
                         # Hence initializing it with vocab_size length
                         n = np.zeros(vocab_size)
                         # Setting the next word to 1 in the one-hot encoded matrix
                         n[word2idx[text.split()[i+1]]] = 1
                         next_words.append(n)

                         images.append(current_image)

                         if count>=batch_size:
                             next_words = np.asarray(next_words)
                             images = np.asarray(images)
                             partial_caps = sequence.pad_sequences(partial_caps, maxlen=max_len, padding='post')
                             yield [[images, partial_caps], next_words]
                             partial_caps = []
                             next_words = []
                             images = []
                             count = 0
```

Let's create the model

FIGURE 14.9 Predicting and generating model.

FIGURE 14.10 Calculating loss and accuracy.

FIGURE 14.11 Beam search for caption search results.

PART II

Soft Computing and Data Analytics

CHAPTER 8

A Heuristic Approach to Parameter Tuning in a SMOTE-Based Preprocessing Algorithm for Imbalanced Ordinal Classification

ANNA SARO VIJENDRAN[1] and D. DHANALAKSHMI[2*]

[1]*School of Computing, Sri Ramakrishna College of Arts and Science, Coimbatore 641006, India*

[2]*Department of Computer Science, Sri Ramakrishna College of Arts and Science, Coimbatore, India*

[*]*Corresponding author. E-mail: dhanadurairaj@gmail.com*

ABSTRACT

Datasets are the backbone for data mining and knowledge engineering field to build an excellent classification model. However, the learning model usually follows a significantly biased distribution of classes. It is known as a class imbalance. The class imbalance problem exists in many real-time datasets. The main objective of this research is to fix an adaptive threshold based on class data and fit that data into models that can be understood and utilized by ordinal multiclass imbalanced scenario for improving the predictive accuracy. The methodology utilizes Laplace-Gauss based SMOTE method for synthesizing sophisticated objects of minority classes. Dynamic parameters are adapted for SMOTE algorithm by utilizing the underlying class information. On the whole, the dataset is divided into training and test data. Training dataset is updated with new synthetic patterns. The experimental analysis is performed on testing dataset to check the efficiency of the proposed methodology by comparing it with the existing methodology. The performance evaluation is conducted in terms of the measures called Mean Absolute Error (MAE),

Maximum Mean Absolute Error (MMAE), Geometric Mean (GM), Kappa, and Average Accuracy. The Experimental results prove that the proposed methodology can produce authentic synthetic patterns than the existing method. The proposed method can synthesize the new effective patterns with the help of dynamic parameter setting. It upgrades the global precision and class-wise precision especially preserves rank order of the classes.

8.1 INTRODUCTION

Machine learning is a subset of artificial intelligence. Although machine learning is a field within computer science, it differs from traditional computational approaches. In traditional computing, algorithms are sets of explicitly programmed instructions used by computers to calculate or solve problems. Machine learning algorithms instead allow for computers to train on data inputs and use statistical analysis in order to get output values that fall within a specific range. Because of this, machine learning facilitates computers in building models from sample data in order to automate decision-making processes based on data inputs. Any technology user today has benefitted from machine learning.

Machine learning is classified into two categories: Supervised learning and unsupervised learning. Classification and regression belong to supervised learning. Clustering is unsupervised learning. From the science perspective, classification is an arrangement of things into an ordered set of related categories based on their closeness and diversity. In machine learning, the classification techniques give more attention to the categorical groups not ordered categorical group. Grouping of objects into ordered related groups is known as ordinal classification.

Ordinal classification resides in between classification and regression. The ordinal classification shows its importance in the following fields. Facial recognition technology to classify the images based on places, time, etc. Recommendation engines suggest movies or television shows watch next based on user preference levels. Automobile engineering field needs the application of ordinal classification in various decision-making processes.

Class imbalance is the main cause to degrade the performance of the classifier. Class imbalance may exist in various flavors, such as small disjuncts, lack of density, borderline patterns, dataset shift, overlapping classes, and noisy data.[1] When class imbalance exists in the ordinal multiclass scenario, it is very difficult for the classifier to learn from the existing available data. The class imbalance occurs either due to the nature of dataset or data that is too expensive or very complicated to acquire such data. Class imbalance yields difficulties in learning to fix the proper decision for such amount of data.

The main types of supervised learning problems include regression and classification problems. Ordinal classification resides in between classification and regression. The ordinal classification shows its importance in the following fields. Facial recognition technology to classify the images based on places, time, etc. Recommendation engines suggest movies or television shows to watch next based on user preference levels. Automobile engineering field needs the application of ordinal classification in various decision-making processes.

Class imbalance is the main cause to degrade the performance of the classifier. When a class imbalance exists in the ordinal multiclass scenario, it is very difficult for the classifier to learn from the existing available data. The nature of class imbalance distribution could occur in two situations, when class imbalance is an intrinsic problem or when it happens naturally. When the data is not naturally imbalanced, it is too expensive to acquire such data for minority class learning due to cost, confidentiality, and tremendous effort to find a well-represented dataset, like a very rare occurrence of the failure of a space shuttle. Class imbalance involves a number of difficulties in learning, including imbalanced class distribution, training sample size, class overlapping, and small disjuncts.

8.1.1 APPROACHES TO TACKLE CLASS IMBALANCE

There are two main approaches for class imbalanced learning: Data level approaches and algorithmic approaches. Data level approaches are oversampling, undersampling, and hybrid sampling. Cost-based learning and ensemble learning belong to algorithmic approaches.

8.1.2 TYPES OF CLASSIFICATION

The binary classification has only two predicted value. More than two predicted values belong to multiclass classification. The ordinal multiclass classification has ordered class labels.

8.2 LITERATURE SURVEY

Synthetic minority oversampling technique (SMOTE) creates new instances in between the seed pattern and one of its k-nearest neighbor. Based on the oversampling rate, patterns are selected from k-nearest neighbors. It makes use of uniform distribution to create new patterns. This is not suitable for complex datasets. It yields overgeneralization.[2] Patterns are categorized into safe, borderline based on near neighbor analysis. New patterns are created at the borderline region. It only takes into account the number of neighbors, so information about prototypes distribution is lost. The standard borderline SMOTE requires a free parameter to find the borderline samples.[3] Safe level SMOTE proposed to overcome the problems associated with borderline SMOTE. This method considers the safe level of the samples on both sides while creating synthetic samples. Based on the safe level ratio, samples are created. It also has some issues, such as the problem of small disjuncts that occurs due to the within-class imbalanced samples. The above-said problem is not tackled well in safe level SMOTE.[4] LN-SMOTE method overcome the issues of safe level SMOTE by considering the local safe level of seed example and one of its k-nearest neighbor. This method has an alternate procedure to calculate the safe level for majority samples and it fixes random value dynamically to create new synthetic examples. The authors also proposed another version of LN-Smote as LN-Smote2. This method uses both oversampling and undersampling. It removes difficult and noisy examples from the majority class, and then applies LN SMOTE.[5] Adasyn algorithm decides the number of synthetic patterns to be created for each selected minority seed pattern.[6] Hybrid preprocessing algorithm proposed with the combination of SMOTE algorithm and rough set theory.[7]

8.2.1 CLASS IMBALANCE IN ORDINAL CLASSIFICATION

Qiao[8] proposed solution for ordinal classification problem through an algorithmic approach. Fouad and Tino[9] suggested two ordinal learning vector quantization classifiers, such as ordinal matrix learning vector quantization and ordinal generalized matrix learning quantization (MLQ). The work makes use of the evaluation metrics, such as mean zero-one-error (MZE), mean absolute error (MAE), maximum mean absolute error (MMAE). QingHua et al.[10] proposed two information entropies in the context of ordinal classification and fuzzy ordinal classification. The work mainly focuses on noisy data. Joao Costa1 et al.[11] used the method called data replication to propose an alternate version of Adaboost. The proposed work is validated using measures, such as mean error rate and MAE. Paul Martin et al.[12] worked on image data to analyze the period of such images using binary before/after classifier. It is one of the applications of the ordinal classification application of dating images. Eibe Frank et al.[13] handled the ordinal classification problem using binary classification problem. Jaime S. Cardoso et al.[14,21] proposed a data replication that which converts the original ordinal problem into two class problem and mapped into support vector machine (SVM) and neural network classifiers. MAE, Spearman and Kendall's tau-b coefficients are used as performance measures.

8.2.2 FINDINGS FROM LITERATURE SURVEY

From the literature review, it is observed that most of the existing work focused on binary classification problem. Due to the complexity of multiclass classification, limited research work carried out in this problem. Likewise, ordinal classification problem handled at algorithmic level. No data level approaches for ordinal classification problem. The literature review suggested that data level approaches are better than algorithmic approaches. Most of the ordinal classification work used the performance measure MAE. The oversampling algorithm use SMOTE-based method for generating new patterns.

8.3 RESEARCH METHODOLOGY

In this chapter, we focus on the issue that the synthetic samples tend to be generated on the line between the minority samples. If there is a significant gap between the majority and the minority, an enormous amount of synthetic data needs to be created. It means that the synthetic data tends to be placed on the same line with high probability. It can be considered as one of the types of over-generalization problem. The proposed method, named as Laplace–Gauss-based SMOTE, can solve the problem by combining Gaussian probability distribution and rectangular probability distribution in the feature space. Both probability distribution can make the SMOTE algorithm to generate new artificial samples deviated in the line but not significantly.

8.3.1 SMOTE

It creates synthetic minority class patterns using feature space similarities. The algorithm picks a minority pattern and finds its k-nearest neighbors and creates synthetic patterns in between the above mentioned two seeds.

For $\forall x_i \in S_{\min}$, one of the k-nearest neighbor $\hat{x}_i$

$X_{new} = x_i + (\hat{x}_i - x_i) \times \delta$ $\quad x_i + (\hat{x}_i - x_i) \times \delta$, where δ is uniformly distributed random variable.

The algorithm synthesizes new patterns until the requiring amount of patterns.

8.3.2 LAPLACE–GAUSS-BASED SMOTE FOR ORDINAL CLASSIFICATION

The SMOTE, Borderline-SMOTE and safe-level-SMOTE algorithms generate synthetic data using uniform probability distribution. However, it is possible to duplicates new patterns. It leads to overfitting.[15] Utilizing the data structure and characteristic is more useful to synthesize new relevant patterns in the feature.[16] We propose a novel Laplace–Gauss SMOTE for imbalanced ordinal classification by combining the two conclusions drawn from.[15,16]

8.3.3 PROPOSED ALGORITHM

- Convert the original data distribution into feature space
- Select the class to be oversampled based on IR value using the following equation:
- $$IR_q = \frac{\sum_{j \neq q} N_j}{Q.N_q} \qquad (8.1)$$
- Calculate the minimum value of the class m_i
- Calculate the maximum value of the class m_x
- Calculate the standard deviation of the class S_d
- Select one random number from uniform probability distribution in between the range of m_i, m_x
- gap $\sim \mathrm{U}(m_i, m_x)$ (8.2)
- Minimum range is calculated using the following equation. Minimum random value is drawn from Gaussian probability distribution in-between gap, S_d
- Minimum range ~ N (gap, S_d) (8.3)
- Maximum random value is drawn from Gaussian probability distribution in-between gap, S_d
- Maximum range ~ N (gap, S_d) (8.4)
- The difference between the minimum range and maximum range is calculated and denoted as the difference.

TABLE 8.1 A Heuristic Approach to Parameter Tuning.

Dataset	Total no. of patterns	No. of attributes	Total no. of classes	IR value per class
Auto	392	7	5	0.65, 0.40, 0.58, 1.14, 7.15
Automobile	205	71	6	12.58, 1.43, 0.33, 0.47, 0.90, 1.11
automobile12vs345vs6	205	71	3	2.50, 0.11, 2.10
Bondrate	57	37	4	1.85, 0.19, 0.92, 2.38
Car	1728	21	4	0.11, 0.88, 5.98, 6.36
ERA	1000	4	9	1.10, 0.68, 0.50, 0.53, 0.60, 0.83, 1.15, 3.51, 5.84
ERA1vs23456vs7vs8vs9	1000	4	5	1.97, 0.06, 2.07, 6.32, 10.51
ESL	488	4	9	20.22, 4.41, 1.29, 0.43, 0.36, 0.29, 0.77, 2.79, 13.44
ESL12vs3vs456vs7vs89	488	4	5	6.45, 2.41, 0.08, 1.39, 3.87
eucalyptus1vs2vs345	736	91	3	1.03, 1.94, 0.21
eucalyptus123vs4vs5	736	91	3	0.25, 0.82, 2.00
heating	768	8	8	4.67, 0.24, 0.73, 1.77,0.68, 1.00, 1.06, 2.64
housing5	506	13	5	1.11, 0.22, 0.62, 2.61, 3.10
LEV	1000	4	5	1.94, 0.51, 0.30, 0.81, 7.30
machine5	209	6	5	0.07, 1.36, 2.92, 6.04, 4.26
machine10	209	6	10	0.08, 0.46, 0.94, 3.02, 2.50, 3.80, 7.70, 5.10, 5.10, 3.80
newthyroid	215	5	3	2.00, 0.15, 1.73
stock	950	9	10	1.88, 0.77, 0.78, 0.70, 0.47, 0.81, 0.81, 0.81, 1.38, 4.35
SWD	1000	4	9	7.56, 0.46, 0.38, 0.90
Toy	300	2	5	1.53, 0.49, 0.56, 0.68, 1.68
triazines5	186	60	5	5.36, 3.27, 1.26, .023, 0.46
triazines10	186	60	10	4.53, 4.53, 13.80, 2.22, 1.64, 1.16, 0.41, 0.28, 0.31, 1.44
wisconsin5	194	32	5	0.38, 0.74, 0.71, 1.14, 1.87
wisconsin10	194	32	10	0.31, 0.81, 0.59, 1.35, 0.71, 1.02, 1.35, 1.71, 1.97, 1.97

- Difference = minimum range − maximum range (8.5)
- The Laplace–Gauss-based SMOTE is given by

$$X_{new} = x_i + (\hat{x}_i - x_i) \times \text{diff} \tag{8.6}$$

8.4 EXPERIMENTAL RESULTS

The first graph-based oversampling algorithm proposed for the context of imbalanced ordinal classification.[17] In this work, they make use of the above-said algorithm to check the performance of the proposed methodology. To validate the proposed methodology some datasets like Wisconsin, housing, stock, machine, triazines, and auto are derived from Chu et al.[17] The rest of the datasets are extracted from UCI. Table 8.1 shows the characteristics of the dataset. Initially, these datasets do not represent ordinal classification, but they represent regression. To evolve this regression into ordinal classification we have considered the desired result is categorized into five or ten classes with equal frequency.

The work considered ELMOP as classifier for both graph-based oversampling and Laplace–Gauss-based SMOTE algorithm.

8.4.1 PERFORMANCE MEASURES

This work preferred most relevant performance measures, such as MAE, MMAE, Geometric mean, Cohen's kappa, accuracy used to validate the proposed work.

8.4.1.1 MEAN ABSOLUTE ERROR

MAE measures the average magnitude of the errors in a set of predictions, without considering their direction[19]

$$MAE_q = \frac{1}{N_q} \sum_{i=1}^{N_q} \left| O(y_i) - O(\hat{y}_i) \right| \tag{8.7}$$

8.4.1.2 MAXIMUM MEAN ABSOLUTE ERROR

MMAE metric for ordinal classification proposed.[20] It displays the maximum MAE for all the classes.

$$\text{MMAE} = \max\left\{MAE_q; q \in \{1, \ldots, Q\}\right\} \tag{8.8}$$

8.4.1.3 GEOMETRIC MEAN

Geometric mean is one of the preferable measures for imbalanced learning. Geometric mean is defined as follows:

$$\text{GMean} = \sqrt{\frac{TP}{TP+FN} \times \frac{TN}{TN+FP}} \quad (8.9)$$

8.4.1.4 KAPPA

Cohen's kappa statistic is one of the preferable measures for imbalanced multiclass learning. When kappa value <0 is indicating no coexists between actual and predicted value, 0–0.20 as slight coexists, 0.21–0.40 as fair coincide, 0.41–0.60 as moderate agreement, 0.61–0.80 as substantial, and 0.81–1 as almost perfect agreement.

$$\text{Kappa} = \frac{N\sum_{i=1}^{i=m} TP - \sum_{i=1}^{i=m} T_{ri}\, T_{ci}}{N^2 - \sum_{i=1}^{i=m} T_{ri}\, T_{ci}} \quad (8.10)$$

where N represents the total number of patterns T_{ri} number of rows from the confusion matrix, and T_{ci}—number of columns from the confusion matrix.

8.4.1.5 ACCURACY

Accuracy is the proportion of true results, either true positive or true negative.

$$\text{Accuracy} = (\text{TP} + \text{TN})/(\text{TN} + \text{TP} + \text{FN} + \text{FP}) \quad (8.11)$$

After evaluating the existing and proposed methodologies with the following measures: MAE, MMAE, GMean, kappa, and accuracy, results are displayed in Tables 8.2–8.6.

TABLE 8.2 MAE.

Dataset	Graph-based oversampling	Laplace–Gauss-based SMOTE
auto	0.4081	**0.3367**
Automobile	0.6153	0.6153
automobile12vs345vs6	0.1346	0.1346
Bondrate	0.6000	**0.1333**
car	0.2453	0.2453
ERA	0.9240	**0.9160**

TABLE 8.2 *(Continued)*

Dataset	Graph-based oversampling	Laplace–Gauss-based SMOTE
ERA1vs23456vs7vs8vs9	0.1320	**0.0920**
ESL	0.4016	0.4016
ESL12vs3vs456vs7vs89	0.1639	**0.1311**
eucalyptus1vs2vs345	**0.4293**	0.4347
eucalyptus123vs4vs5	**0.2391**	0.3315
heating	0.7604	0.7604
housing5	**0.1102**	0.1653
LEV	0.3760	0.3760
machine5	**0.2641**	0.3018
machine10	**0.4528**	0.6226
newthyroid	0.0740	**0.0555**
stock	**0.9873**	1.0210
SWD	0.2680	0.2680
toy	**0.4800**	0.5866
triazines5	0.3404	0.3404
triazines10	0.7021	0.7021
wisconsin5	0.3469	0.3469
wisconsin10	1.1837	**1.0204**

The values displayed in bold faces signify the improvements of performance than the existing Graph-based algorithm.

TABLE 8.3 MMAE.

Dataset	Graph-based oversampling	Laplace–Gauss-based SMOTE
Auto	1.0434	**1.0000**
Automobile	3.0000	3.0000
automobile12vs345vs6	1.0000	1.0000
Bondrate	1.5000	**1.0000**
Car	0.3498	0.3498
ERA	3.5217	**3.2608**
ERA1vs23456vs7vs8vs9	1.0434	**1.0000**
ESL	1.2222	1.2222
ESL12vs3vs456vs7vs89	2.3333	**1.0000**
eucalyptus1vs2vs345	1.2000	1.2000
eucalyptus123vs4vs5	**0.4190**	0.5809
Heating	1.8358	1.8358
housing5	**0.3684**	1.0000
LEV	1.5217	1.5217
machine5	**0.3684**	0.4210

TABLE 8.3 *(Continued)*

machine10	**0.8275**	1.1379
Newthyroid	0.1428	0.1428
Stock	4.2500	4.2500
SWD	1.5000	1.5000
Toy	**1.7500**	2.2500
triazines5	3.0000	3.0000
triazines10	6.0000	6.0000
wisconsin5	1.0000	1.0000
wisconsin10	3.1667	**2.7500**

The values displayed in bold faces signify the improvements of performance than the existing Graph-based algorithm.

TABLE 8.4 Accuracy.

Dataset	**Graph-based oversampling**	**Laplace–Gauss-based SMOTE**
Auto	0.6531	**0.7143**
Automobile	0.5385	0.5385
automobile12vs345vs6	0.8654	0.8654
Bondrate	0.5333	**0.8667**
Car	0.8773	0.8773
ERA	0.5440	**0.5480**
ERA1vs23456vs7vs8vs9	0.872	**0.9080**
ESL	0.6777	0.6777
ESL12vs3vs456vs7vs89	0.8689	**0.8770**
eucalyptus1vs2vs345	**0.7174**	0.7120
eucalyptus123vs4vs5	**0.7609**	0.6685
Heating	0.5625	0.5625
housing5	**0.8898**	0.8346
LEV	0.6720	0.6720
machine5	0.8491	0.8491
machine10	**0.8868**	0.7358
Newthyroid	0.9259	**0.9444**
Stock	**0.5924**	0.5882
SWD	0.7480	0.7480
Toy	**0.6000**	0.5867
triazines5	0.7660	0.7660
triazines10	0.3683	0.3683
wisconsin5	0.6531	0.6531
wisconsin10	0.5306	**0.5510**

The values displayed in bold faces signify the improvements of performance than the existing Graph-based algorithm.

TABLE 8.5 GMean.

Dataset	Graph-based oversampling	Laplace–Gauss-based SMOTE
Auto	0.8129	**0.8403**
Automobile	0.6688	0.6688
automobile12vs345vs6	0.7396	0.7396
Bondrate	**0.6884**	0.8292
Car	0.9622	0.9622
ERA	0.7774	**0.7792**
ERA1vs23456vs7vs8vs9	0.8501	0.8576
ESL	0.8052	0.8052
ESL12vs3vs456vs7vs89	0.7663	**0.8067**
eucalyptus1vs2vs345	**0.6020**	0.5921
eucalyptus123vs4vs5	**0.8740**	0.8253
heating	0.8663	0.8663
housing5	**0.9332**	0.8631
LEV	0.7556	0.7556
machine5	0.9626	**0.9629**
machine10	**0.9836**	0.9611
newthyroid	0.9417	**0.9496**
Stock	**0.7555**	0.7539
SWD	0.7310	0.7310
Toy	**0.7310**	0.7240
triazines5	0.6051	0.6051
triazines10	0.5951	0.5951
wisconsin5	0.8535	0.8535
wisconsin10	0.8325	**0.8401**

The values displayed in bold faces signify the improvements of performance than the existing Graph-based algorithm.

TABLE 8.6 Kappa.

Dataset	Graph-based oversampling	Laplace–Gauss based SMOTE
Auto	0.0776	**0.1071**
Automobile	0.3981	0.3981
automobile12vs345vs6	0.6971	0.6971
Bondrate	0.2969	**0.6444**
car	0.6728	0.6728
ERA	**0.5668**	0.5630

TABLE 8.6 *(Continued)*

Dataset	Graph-based oversampling	Laplace–Gauss based SMOTE
ERA1vs23456vs7vs8vs9	0.6000	**0.7125**
ESL	0.3213	0.3213
ESL12vs3vs456vs7vs89	0.5902	**0.6158**
eucalyptus1vs2vs345	**0.3641**	0.3519
eucalyptus123vs4vs5	**0.4620**	0.2541
heating	0.5000	0.5000
housing5	**0.6555**	0.4833
LEV	0.0244	0.0244
machine5	0.5283	0.5283
machine10	**0.3711**	0.3186
newthyroid	0.8333	**0.8750**
stock	0.5584	**0.5629**
SWD	0.3280	0.3280
toy	0.2000	**0.2258**
triazines5	0.2686	0.2686
triazines10	0.4539	0.4539
wisconsin5	0.0776	0.0776
wisconsin10	**0.6165**	0.5991

The values displayed in bold faces signify the improvements of performance than the existing Graph-based algorithm.

8.5 ANALYSIS OF RESULTS

8.5.1 MEAN RANKING OF PERFORMANCE MEASURES

Based on the above results we can conclude some interesting observations. Table 8.2 shows both existing and proposed methods perform equally. MMAE is the measure exclusively for ordinal classification. Table 8.3 indicates Laplace–Gauss-based SMOTE outperforms than graph-based oversampling. Tables 8.4–8.6 prove that the proposed method performs better than an existing method for measuring accuracy, GMean, and kappa. Figure 8.1 displays the mean ranking of the results obtained from Tables 8.2–8.6 of both methods.

8.5.2 STATISTICAL ANALYSIS

The *t*-test is performed on the mean ranking of the evaluation measures of the existing method and proposed a method to find the statistical significance of the heuristic approach of parameter tuning. Paired two samples for means is used on performance metrics such as MAE, MMAE, accuracy, GMean, and kappa to test twice. The usual reason for performing this test is when you are testing the same group twice to compare the results before and after the parameter tuning.

To quantify whether a statistical difference exists among the algorithms compared, *t*-test is performed on the mean ranking of all the evaluation measures it is displayed in Table 8.7.

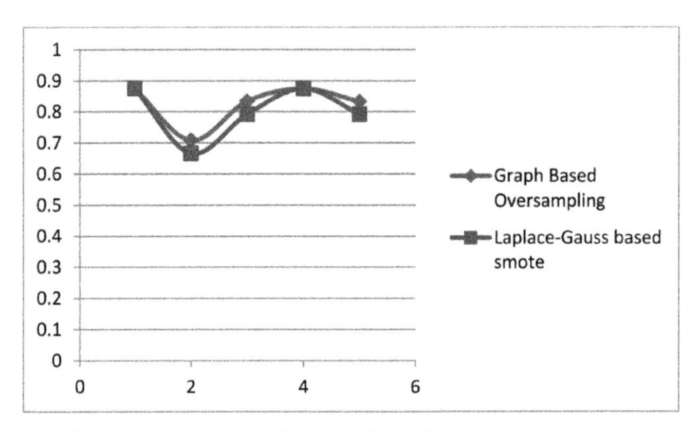

FIGURE 8.1 (See color insert.) Mean ranking of performance measures.

TABLE 8.7 *t*-Test Paired Two Sample for Means on Mean Ranking of the Evaluation Measures ($\alpha = 0.05$).

	Variable 1	Variable 2
Mean	0.8332	0.7912
Variance	0.006111	0.006111
Observations	5	5
Pearson correlation	0.927837	
Hypothesized mean difference	0	
Df	4	
t Stat	3.162278	
$P(T \leq t)$ one-tail	0.017055	
t Critical one-tail	2.131847	
$P(T \leq t)$ two-tail	0.034109	
t Critical two-tail	2.776445	

The above results indicate that the p values of one tail and two tail are less than the alpha value (0.05). In addition to that the t value is larger than the t critical values of one tail and two tail. The above said p value and t values prove that the null hypothesis is rejected and the two methods perform more or less similarly.

8.5.3 MEAN RANKING OF EXISTING AND PROPOSED METHOD

The Graph-based oversampling and Laplace–Gauss-based SMOTE evaluated on 24 datasets with five evaluation measures. Table 8.8 shows the mean ranking of the above said two methods based on the best results of all five evaluation measure.

TABLE 8.8 Mean Rank.

Method	Rank
Graph-based oversampling	0.46
Laplace–Gauss-based SMOTE	0.42

Figure 8.2 depicts that, Laplace–Gauss based SMOTE performs better than Graph-based oversampling method with ELMOP as a classifier.

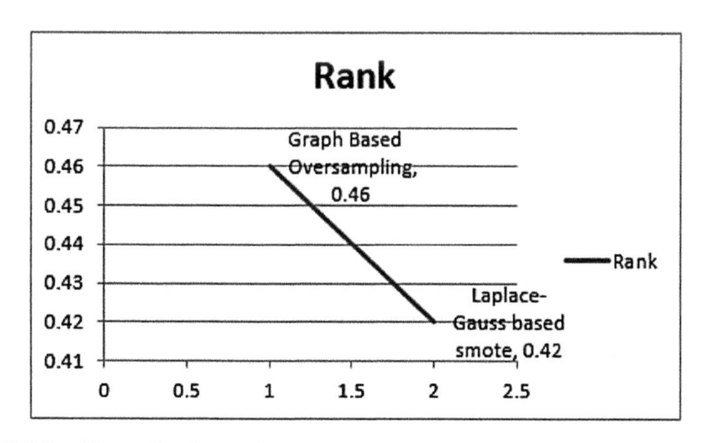

FIGURE 8.2 **(See color insert.)** Mean ranking of two methodologies.

8.6 CONCLUSIONS AND FUTURE WORK

Datasets are the backbone for data mining and knowledge engineering field to build an excellent classification model. However, the learning model usually follows a significantly biased distribution of classes. It is known as class imbalance. The class imbalance problem exists in many real-time datasets. The main objective of this research is to fix adaptive threshold based on class data and fit that data into models that can be understood and utilized by ordinal multiclass imbalanced scenario for improving the predictive accuracy. The methodology utilizes Laplace–Gauss based SMOTE method for synthesizing sophisticated objects of minority classes. Dynamic parameters are adapted for SMOTE algorithm by utilizing the underlying class information. On the whole, the dataset is divided into training and test data. Training dataset is updated with new synthetic patterns. The experimental analysis is performed on testing dataset to check the efficiency of the proposed methodology by comparing it with the existing methodology. The performance evaluation is conducted in terms of the measures called MAE, MMAE, geometric mean, kappa, and average accuracy. The experimental results prove that the proposed methodology can produce better authentic synthetic patterns than the existing method. The proposed method can synthesize the new effective patterns with the help of dynamic parameter setting. It upgrades the global precision and class-wise precision especially preserving rank order of the classes.

Acquiring learning patterns from the real world has lots of problems including noise, confidentiality, etc., which may lead to class imbalance nature or skewed class distribution. In this chapter, we have discussed the problem of learning from imbalanced ordinal multiclass datasets. Such scenarios give significantly great challenge than binary imbalanced problems. Therefore, it is important to resolve such a complicated the problem. In this chapter, we proposed a novel method named Laplace–Gauss-based SMOTE for imbalanced ordinal multiclass by combining Gaussian probability distribution and rectangular probability distribution in the feature space. In addition to this, we propose to base the preprocessing on extracted class information to heuristically tune the threshold for SMOTE algorithm.

The experimental results prove that the proposed methodology yields better performance than the Graph-based Oversampling by using the benchmark datasets. For future work, we will discard noisy data to improve performance and reduce the error.

KEYWORDS

- **multiclass ordinal classification**
- **heuristics**
- **extreme learning machine for ordinal regression**
- **maximum mean absolute error**
- **geometric mean**
- **Kappa**
- **average accuracy**

REFERENCES

1. Lopez, V.; Fernandez, A.; Garcia, S.; Palade, V.; Herrera, F. An Insight into Classification with Imbalanced Data: Empirical Results and Current Trends on Using Data Intrinsic Characteristics. *Inform. Sci.* **2013,** *250*, 113–141.
2. Chawla, N.; Bowyer. K.; Hall, L.; Kegelmeyer, W.. SMOTE: Synthetic Minority Over-sampling Technique. *J. Artif. Intell. Res.* **2002,** *16*, 341–378.
3. Han, H.; Wang, W. Y.; Mao, B. H.; In *Borderline SMOTE: A New Over-Sampling Method in Imbalanced Data Sets Learning,* Proceedings of the International Conference on Intelligent Computing, Part I, LNCS, 2005, 3644, pp 878–887.
4. Bunkhumpornpat, C.; Sinapiromsaran, K.; Lursinsap, C. *Safe-Level-SMOTE: Safe Level Synthetic Over-Sampling Technique for Handling the Class Imbalanced Problem*, Proceedings of the PAKDD 2009, Springer LNAI, 5476, 2009, pp 475–482.
5. Maciejewski, T.; Stefanowski, J.; Local Neighbourhood Extension of SMOTE for Mining Imbalanced Data, IEEE Symposium on Computational Intelligence and Data Mining (CIDM), Paris, France, 2011. DOI: 10.1109/CIDM.2011.5949434
6. He, H.; Bai, Y.; Garcia, E. A.; Li, S. *ADASYN: Adaptive Synthetic Sampling Approach for Imbalanced Learning*, Proceedings of IEEE International Joint Conference on Neural Networks, Hong Kong, 2008, pp 1322–1328.
7. Ramentol, E.; Caballero, Y.; Bello, R.; Herrera, F. Smote-RSB: A Hybrid Preprocessing Approach Based on Oversampling and Undersampling for High Imbalanced Data-Sets Using Smote and Rough Sets Theory. *KnowledgeInform. Syst.* **2012,** *33* (2), 245–265.
8. Qiao, X. Non-crossing Ordinal Classification, Statistics and Its Interface arXiv, Cornell University: USA, 2015
9. Fouad, S.; Tino, P. Adaptive Metric Learning Vector Quantization for Ordinal Classification" Downloaded from Internet. *Neural Comput.* **2012,** *24* (11), 2825–2851.
10. QingHua, H. U.; MaoZu, G.; DaRen, Y.; JinFu, L. Information Entropy for Ordinal Classification. *Sci. China Inform. Sci.* **2010,** *53* (6), 1188–1200.

11. Costal, J.; Cardoso, J. S. *oADABOOST-An ADABOOST Variant for Ordinal Classification", Downloaded from Internet,* International Conference on Pattern Recognition Applications and Methods, 2015.
12. Martin, P.; Doucet, A.; Jurie, F. *Dating Color Images with Ordinal Classification,* ICMR'14, Glasgow, United Kingdom, 2014.
13. Frank, E.; Hall, M. *A Simple Approach to Ordinal Classification,* European Conference on Machine Learning, 2001, pp 145–156.
14. Cardoso, J. S.; Costa, J. F. P.; Learning to Classify Ordinal Data: The Data Replication Method. *J. Mach. Learn. Res.* **2007,** *8,* 1393–1429.
15. Lee, H.; Kim, J.; Kim, S. Gaussian-Based SMOTE Algorithm for Solving Skewed Class Distributions. *Int. J. Fuzzy Logic Intell. Syst.* **2017,** *17* (4), 229–234.
16. Saez, J. A.; Krawczyk, B.; Wozniak, M. Analysisng the Oversampling of Different Classes and Types of Examples in Multi-class Imbalanced Dataset. *Pattern Recog.* **2016,** *57,* 164–178.
17. Perez-Ortiz, M.; Gutierrez, P. A.; Hervas-Martınez, C.; Yao, X.; Graph Based Approaches for Over-sampling in the Context of Ordinal Regression. *IEEE Trans. Knowl. Data Eng.* **2015,** *27* (5), 1233–1245.
18. Chu, W.; Ghahramani, Z. Gaussian Processes for Ordinal Regression. *J. Mach. Learn. Res.* **2005,** *6,* 1019–1041.
19. Willmott, C. J.; Matsuura, K. Advantages of the Mean Absolute Error (MAE) Over the Root Mean Square Error (RMSE) in Assessing Average Model Performance. *Clim. Res.* **2005,** *30,* 79–82.
20. Cruz-Ramırez, M.; Hervas-Martınez, C.; Sanchez-Monedero, J.; Gutierrez, P. A. Metrics to Guide a Multi-objective Evolutionary Algorithm for Ordinal classification. *Neurocomputing* **2014,** *135,* 21–31.
21. Chowdhary, C. L.; Acharjya, D. P. Segmentation of Mammograms Using a Novel Intuitionistic Possibilistic Fuzzy C-Mean Clustering Algorithm. In *Nature Inspired Computing;* Springer: Singapore, 2018; pp 75–82.

CHAPTER 9

Aspects of Deep Learning: Hyper-Parameter Tuning, Regularization, and Normalization

SYED MUZAMIL BASHA[1*] and DHARMENDRA SINGH RAJPUT[2]

[1]*Department of Information Technology, Sri Krishna College of Engineering and Technology, Coimbatore 641008, Tamil Nadu, India*

[2]*School of Information Technology and Engineering, Vellore Institute of Technology, Vellore, India*

[*]*Corresponding author. E-mail: muza.basha@gmail.com*

ABSTRACT

Deep learning today is applied to many different application areas and that intuitions about hyper-parameter settings from one application will differ to another application. There is a lot of cross-fertilization among different application domains. To implement more complex models, such as convolutional neural networks or recurring neural networks, is not practical to implement everything yourself from scratch. Fortunately, there are now many good deep learning software frameworks that can help you implement these models. This chapter, focus on learning how to implement a Neural Network. Ranging from aspects like: Hyper-parameter tuning helps the network to train quickly. Setting up user data helps in quickly finding a good high-performance neural network. Analyzing the impact of bias and variance on overall performance. Applying different forms of regularization toward reducing variance of Neural network. Making Optimization algorithm runs quickly with reasonable learning rate using Batch normalization. Additionally, A sense of the typical structure of a TensorFlow program is developed using Python in Jupyter notebook and to start up TensorFlow. The Recommendations made by the research work is as follows: 99.5% training and 0.25% development, 0.25% testing dataset ratio, is preferred in setting up the data. L2 Regularization is used to reduce variance, Adam optimization algorithm to be used update weights of the gradient. Batch normalization makes hyper-parameter search problem much easier, makes the neural network much more robust. So all the modern deep learning programming framework makes it really easy to code up even pretty complex neural networks.

9.1 INTRODUCTION

In this chapter, we aim to make the learners understand implementing a neural network. The practical aspects of neural network are: hyper-parameter tuning, setting up data, optimization algorithm to train the model. When coming up with a replacement application, it is virtually not possible to properly guess the correct values for all of those on the primary try. In follow, applied machine learning could be an extremely repetitive method, which frequently starts with a plan. Such as, to make a neural network of an exact range of layers, the bound range of hidden units, perhaps on specific information sets and then on. Supported the end result, one would possibly refine the ideas and alter the alternatives created to search out a far better neural network.

Today, deep learning has found nice success in several areas: natural language processing (NLP) (Basha, 2018), Speech recognition and heaps of applications on structured information (Basha, 2017). Structured information includes everything from advertisements to net search. The choice taken in planning depends on the quantity of knowledge and variety of input options. It was nearly not possible to properly guess the simplest alternative of hyper-parameters the initial time. Applied deep learning could be a terribly reiterative method, wherever one ought to have to be compelled to go around this cycle persistently to find an honest alternative of network for every application. To determine how quickly and how efficiently one can go around this cycle helps in building an unbiased network. The chapter is organized as follows: discussion on setting up user data helps in quickly finding a good high-performance neural network, analyzing the impact of bias and variance on overall performance, applying different forms of regularization toward reducing variance of neural network, making optimization algorithm dashes with reasonable learning rate using batch normalization, discussion on open source programming frameworks of deep learning (TensorFlow). Additionally, a sense of the typical structure of a TensorFlow program is developed using Python in Jupyter notebook and to start up TensorFlow.

Deep learning today is applied to many different application areas and that intuitions about hyper-parameter settings from one application will differ to another application. There is a lot of cross-fertilization among different application domains. To implement more complex models, such as convolutional neural networks or recurring neural networks, is not practical to implement everything yourself from scratch. Fortunately, there are now many good deep learning software frameworks that can help you implement these models. This chapter, focus on learning how to implement a neural network. The aspects covered in this chapter are: hyper-parameter tuning helps the network to train quickly. Setting up user data helps in quickly finding a good high-performance neural network, analyzing the impact of bias and variance on overall performance, and applying different forms of regularization toward reducing variance of neural network. Making Optimization algorithm runs quickly with reasonable learning rate using batch normalization. Additionally, a sense of the typical structure of a TensorFlow program is developed using Python in Jupyter notebook and to start up TensorFlow. The Recommendations made by the research work is as follows: 99.5% training and 0.25% development, 0.25% testing dataset ratio is preferred in setting up the data. L2 regularization is used to reduce variance, Adam optimization algorithm to be used to update weights of gradient. Batch normalization makes hyper-parameter search problem much easier, makes neural network much more robust. So all the modern deep learning programming framework makes it really easy to code up even pretty complex neural networks.

9.2 METHODOLOGY

In setting up, datasets well relating to training, development and testing sets will build rather more economical. Traditionally, one would possibly take all the information (or) carve off some portion of it to be training set. Some portion of it to be hold-out cross-validation set, and this can be referred to as the development set. The final portion of it to be tested, trained and developing the dataset. This helps to estimate the performance of the machine learning models. Therefore, the previous era of machine learning, it had been common observation to require all knowledge and split it consistent with perhaps a 70–30% in train-test splits. If one contains a specific development set, split regarding 60% train, 20% dev, and 20% test datasets. The goal of the development (dev) set is to test totally different algorithmic programs thereon and notice that algorithm works higher. So, the dev set must be large enough to judge, totally different algorithmic program selections and quickly decide that one is doing higher. Even over 1,000,000 examples, would possibly find 99.5% train and 0.25% dev, 0.25% test (or) maybe a 0.4% dev, 0.1% test (Levine, 2018). A comparatively tiny dataset, these ancient ratios could be sensible in the machine learning world, once has train and dev set however no separate test set, truly find yourself doing is mistreatment the test set as a hold-out cross-validation set. The matter with hold-out cross-validation is over-fitting to the test set. So, deciding the ratios of a train, dev and test set, permits in evaluating the efficiency of any machine learning algorithm.

In the deep learning, error referred to as the bias–variance trade-off (Glaze, 2018). Whereas training a neural network, have trained associate initial model. Check whether or not the algorithmic rule contains a high bias? If affirmative, add additional hidden layers or additional hidden units (or) attempt additional advanced optimization algorithms (or) realize a replacement spec that is higher suited to this drawback. Best thanks to solve a high variance drawback is to urge additional information. However, obtaining additional information when is tough (or) use regularization, to cut back over-fitting (Peng, 2018). Therefore, for instance, having a high bias drawback, obtaining additional training information is not reaching to facilitate (or) a minimum of it is not the foremost economical issue to try. Therefore, being clear on what quantity of a bias drawback or variance drawback or each will facilitate in specializing in choosing the foremost helpful things to do. Second, within the earlier era of machine learning, there accustomed be abundant discussion on what is referred to as the bias–variance exchange. Moreover, the explanation for that was that one may increase bias and cut back variance, or cut back bias and increase variance. we tend to state that this has one in every of the large reasons that deep learning has been therefore helpful for supervised learning, that there is abundant less of this exchange wherever you have got to balance bias and variance fastidiously, however typically having additional choices for reducing bias or reducing variance while not essentially increasing the opposite one. The essential structure of the way to organize any machine learning drawback to diagnose bias and variance, and so attempt to choose the proper operation for creating progress on any drawback. Regularization could be a helpful technique for reducing variance. Now, allow us to have associate understanding of the way to apply regularization to any neural network.

If a neural network is overfitting the data (high variance problem), first try with regularization technique. The other way to address high variance is to get more training data that is also quite reliable. However, it could be expensive to get more data. Whereas, adding regularization help to prevent over-fitting (or) to reduce the errors in any network. Now, let us learn how regularization works using logistic regression (Jiang, 2018) the cost function is as follows in eq 9.1:

$$J(\theta) = \frac{1}{2m}\left[\sum_{i=1}^{m}\left(E_{\theta}\left(x^{i}\right) - y^{i}\right)^{2} + \lambda\sum_{i=1}^{n}\theta_{j}^{2}\right]. \tag{9.1}$$

where the term $\lambda\sum_{i=1}^{n}\theta_j^2$ should be penalized to reduce the high variance problem, by choosing an extremely large regularization parameter (λ). That in turns led to under-fitting the data that is addressed by carefully selecting the regularization parameter using cross-validation. In which, the values are tested for the best fit of data. Machine learning process comprises several different steps. One, an algorithm to optimize the cost function J using various tools such as grade intersect, momentum, RMS prop, and atom (Aljarah, 2018).

9.3 MINI-BATCH GRADIENT DESCENT

Vectorization allows to efficiently compute on m samples (Li, 2018). Let us take training examples and stack them into the massive matrix of $X = \{X_1, X_2, X_3 \ldots X_M\}$. Moreover, similarly for $Y = \{Y_1, Y_2, Y_3 \ldots Y_M\}$. The dimension of X is $X \times M$ and Y is $1 \times M$. Where the Mini-batch t is represented as in eq 9.2.

$$t = X^{\{t\}}, Y^{\{t\}} \tag{9.2}$$

When the M is huge say 5,000,000. Then using this vectorization approach, one can represent 5000 mini-batches or 1000 each as in eq 9.3.

$$X^{\{1\}} = \{X^{(1)}, X^{(2)}, X^{(3)}, \ldots X^{(1000)}\} \tag{9.3}$$

In particle, Mini-batch gradient descent can be implemented as in algorithm (9.1).

Algorithm 9.1: Mini-Batch Gradient Descent

Step 1: Start
Step 2: for $t = 1\ldots 5000$
Step 3: for each $X^{\{t\}}$

$$Z^{[1]} = W^{[1]} * X^{\{t\}} + b^{[1]}$$
$$A^{[L]} = g^{[lL}(Z^{[L]})$$

Step 4: Compute the cost of J as in eq 9.1.
Step 5: Implement backpropagation to compute gradient concerning $X^{\{t\}}, Y^{\{t\}}$
Step 6: $W^{[L]} = W^{\{L\}} - \alpha \times dW^{[L]}$
Step 7: $b^{[L]} = b^{\{L\}} - \alpha \times db^{[L]}$
Step 8: stop

The algorithm is for one epoch of training and epoch is a word that means a single pass through the training set. Whereas with batch gradient descent, a single pass through the training allows taking only one gradient descent step, with mini-batch gradient descent, a single pass through the training set, that is one epoch, allows you to take 5000 gradient descent steps. So, keep taking passes through the training set until the result converges with approximately converge. When one has a lost training set, mini-batch gradient descent runs much faster than batch gradient descent (Basha, 2018; Chowdhary and Acharjya, 2018).

In exponentially weighted average, Initially, $V_{(0)} = 0$, $W = 0.9$ for t observations the expected value can be computed as in eq 9.5.

$$V_{(t)} = W \bullet V_{(t-1)} + (1 - W) \bullet X_{(t)} \tag{9.4}$$

When $W = 0.9$: approximately 10 values are considered
$W = 0.98$: approximately 50 values are considered
$W = 0.5$: approximately 2 values are considered
Where approximation of $V_{(t)}$ is computed as in eq 9.5.

$$V_{(t)} \cong \frac{1}{1 - W} \tag{9.5}$$

So one of the advantages of this exponentially weighted average is that it takes very little memory and you keep on overwriting it with the latest value (Tran, 2018). However, to compute a moving window, where explicitly sum over the last 10 days, the last 50 days' temperature and divide by 10 or divide by 50, which usually give the better estimation. However, the disadvantage of explicitly keeping all the temperatures around and some of the last 10 days is it requires more memory and it is just more complicated to implement, and is computationally more expensive.

Whereas, gradient descent with the momentum that works quicker than the quality gradient descent algorithmic program. The essential plan is to work out associate in nursing exponentially weighted average of gradients, and so use that gradient to update weights (Srinivasan, 2018). In gradient, descents take several steps, slowly oscillate toward the minimum. This oscillation slows down gradient descent and prevents from employing an abundant larger learning rate. To do so, compute *dw* and *db* on current Mini-Batch as in eq 9.6.

$$\begin{aligned} V_{dw} &= \beta V_{dw} + (1-\beta)dw \\ V_{db} &= \beta V_{db} + (1-\beta)db \end{aligned} \quad (9.6)$$

where $w = w - \alpha V_{dw}, b = b - \alpha V_{db}$

Root Mean Square prop (RMS prop) that can also speed up gradient descent (Paul, 2018). The computation is as shown in eq 9.7.

$$S_{dw} = \beta \times S_{dw} + (1-\beta) \times dw^2$$

$$S_{db} = \beta \times S_{db} + (1-\beta) \times db^2$$

$$\text{Where } w = w - \alpha \times \frac{dw}{\sqrt{S_{dw}}}, b = b - \alpha \times \frac{db}{\sqrt{S_{db}}} \quad (9.7)$$

Recall that in the horizontal direction *w* learning to go faster. Whereas in the vertical direction *b* wants to slow down all the oscillations into the vertical direction. So with this terms, S_{dW} is a relatively small number, and S_{db} the relatively large number, to slow down the updates on a vertical dimension. These derivatives are much more significant in the vertical direction than in the horizontal direction.

9.4 ADAM OPTIMIZATION ALGORITHM

Adam optimization algorithm is implemented by combining momentum and RMS prop (Jørgensen, 2018) as in algorithm (9.2).

Algorithm 9.2: Adam optimization algorithm

Step 1: Start

Step 2: initiate $V_{dw} = 0, S_{dw} = 0, V_{db} = 0, S_{db} = 0$

Step 3: On every iteration *t*: compute dw, db using current Mini-Batch

Step 4: $V_{dw} = \beta_1 V_{dw} + (1-\beta_1)dw$, $V_{db} = \beta_1 V_{db} + (1-\beta_1)db : \beta_1 = Momentum$

Step 5: $S_{dw} = \beta_2 S_{dw} + (1-\beta_2)dw^2$, $S_{db} = \beta_2 S_{db} + (1-\beta_2)db : \beta_2 = RMS\ prop$

Step 6: $V_{dw}^{c} = {V_{dw}}/{(1-\beta_1^t)}$, $V_{db}^{c} = {V_{db}}/{(1-\beta_1^t)}$

Step 7: $S_{dw}^{c} = {S_{dw}}/{(1-\beta_2^t)}$, $S_{db}^{c} = {S_{db}}/{(1-\beta_2^t)}$

Step 8: update the weights as

$$w := w - \alpha \times {V_{dw}^{c}}/{\sqrt{S_{dw}^{c} + \varepsilon}}, \ b := b - \alpha \times {V_{db}^{c}}/{\sqrt{S_{db}^{c} + \varepsilon}}$$

Step 9: stop

A default choice for β_1 is 0.9 (moving average, the weighted average of *dw* right this is the light momentum term). The hyper-parameter for β_2 is 0.999. Next, setting up any optimization problem to make training go quickly. When training a neural network, one of the techniques that speed up training is by normalizing inputs. Normalizing inputs corresponds to two steps. The first is to subtract out or to zero out the mean. Moreover, then the second step is to normalize the variances. In particular, one does not want to normalize the training set and the test set differently. One thing that makes it more difficult is that Deep Learning does not work best in a regime of big data. So, what to find is that having fast optimization algorithms, having good optimization algorithms can speed up the efficiency of the team. One of the things that might help speed up any learning algorithm is to slowly reduce the learning rate over time termed as learning rate decay. To speed up any learning algorithm, slowly reduce learning rate over time. These are termed as learning rate decay. The Learning rate decay can be addressed as shown in the eq 9.8.

$$\alpha = \frac{1}{1 + decay\_rate \times epoch\_number} \times \alpha_0 \tag{9.8}$$

where $\alpha_0 = 0.2$, $decay\_rate = 1$ the learning rate can be gradually decreased as in Table 9.1.

TABLE 9.1 Learning Rate Decay.

Epoch_Number	α
1	0.1
2	0.67
3	0.5
4	0.4

In the early days of deep learning, people used to worry a lot about the optimization algorithm getting stuck in bad local optima (Laurent, 2018). It is essential to pick the appropriate scale to explore the hyper-parameters. Consider the range (r) of values as shown in eq 9.9.

$$r = -4 \times rand(), \quad r \in [-4, 0]$$
$$\alpha = 10^r, \quad \alpha \in [10^{-4}, ..., 10^0]$$
$$a = \log_{10}^{0.0001} = -4, \quad b = \log_{10}^{1} = 0, \quad r \in [a, b] \tag{9.9}$$

To sample r uniformly at random between a and b, r would be between −4 and 0. To set α, on randomly sampled hyper-parameter value, as in eq 9.9. Finally, one other tricky case is sampling the hyper-parameter β, used for computing exponentially weighted averages. Let us consider that, β should be between 0.9 and 0.999. When computing exponentially weighted averages, using 0.9 is like averaging over the last 10 values. Whereas using 0.999 is like averaging over the last 1000 values as shown in eq 9.10.

$$\beta = \{0.9, ..., 0.999\}$$
$$1 - \beta = \{0.1, ..., 0.001\}$$

$$1 - \beta = 10^r$$
$$\beta = 1 - 10^r \tag{9.10}$$

When β is close to 1, the sensitivity of the results changes even with minimal changes in β. So, if β goes from 0.9 to 0.9005, there is hardly any change in the results. However, if β goes from 0.999 to 0.9995, this has a significant impact on the performance of the algorithm. In both of these cases, it is averaging over roughly 10 values. However, from an exponentially weighted average over about the last 1000 examples, to now, the last 2000 examples. Moreover, it is because that formula is, $\frac{1}{1-\beta}$, this is very sensitive to small changes in the beta, when the beta is close to 1.

Deep learning today is applied to many different application areas, and that intuitions about hyper-parameter settings from one application differ to another application. There is much cross-fertilization among different application domains. There are two approaches to set the value of hyper-parameters (Pham, 2018): one approach, babysit one model that is watching the performance and patiently nudging the learning rate up or down. However, what happens if one does not have enough computational capacity to train many models at the same time. The other approach would be if you train many models in parallel. It might have some set of the hyper-parameters and just let it run by itself, either for a day or even for multiple days. Moreover, then at the same time, start up a different model with a different set of the hyper-parameters. Training many different models in parallel, a lot of different hyper-parameter settings and then just maybe quickly at the end pick the one that works best. So, to make an analogy, call the first approach as panda approach. When pandas have children, they have very few children, usually one child at a time, and then they put much effort into making sure that the baby panda survives. So that is babysitting. One model or one baby panda. Whereas, the second approach is more like what fish do, call this the caviar strategy. There is some fish that lay over 100 million eggs in one mating season. However, the way fish reproduce is they lay many eggs and do not pay too much attention to anyone of them but see that hopefully one of them, or maybe a bunch of them, perform well. So the way to choose between these two approaches is a function of how much computational resources you have. If you have enough computers to train many models in parallel, then, by all means, take the caviar approach and try a lot of different hyper-parameters and see what works. However, in some application domains, I see this in some online advertising settings as well as in some computer vision applications, where there is just so much data, and the models you want to train are so big that it is difficult to train many models at the same time. It is application dependent. Communities use the panda approach a little bit more, we are babying a single model along and nudging the parameters up and down and trying to make this one model work.

9.5 BATCH NORMALIZATION

Batch normalization makes hyper-parameter search problem much more comfortable, makes neural network much more robust (Li, 2018). The choice of hyper-parameters is a much bigger range of hyper-parameters that work well, and also enable to much more easily train even very deep networks. In the case of logistic regression, we saw how normalizing x_1, x_2, x_3 helps train w and b more efficiently as stated in eq 9.11.

$$\mu = \frac{1}{m}\sum_{i=1}^{n} X^{(i)}$$
$$X = X - \mu$$
$$\sigma^2 = \frac{1}{m}\sum_{i=1}^{n} (X^{(i)})^2 \tag{9.11}$$
$$X = X - \sigma^2$$

For the hidden layers in the network, consider the intermediate values as $Z = \{Z^{(1)}, \ldots, Z^{(m)}\}$. The intermediate value can be represented as $Z^{[l](i)}$ the normalized $Z^{[l](i)}$ can be computed as in eq 9.12.

$$\mu = \frac{1}{m}\sum_{i=1}^{n} Z^{[l](i)}$$
$$\sigma^2 = \frac{1}{m}\sum_{i=1}^{n} (Z^{[l](i)} - \mu)^2 \tag{9.13}$$
$$Z_{norm}^{[l](i)} = \frac{Z^{[l](i)} - \mu}{\sqrt{\sigma^2 + \varepsilon}}$$
$$Z^{N(i)} = \chi \times Z^{[l](i)} + \beta$$

Where χ and β are learnable parameters of the model. Using gradient descent (or) gradient descent of momentum, (or) RMS prob, (or) Atom, updates the parameters χ and β. Now, notice that the effect of χ and β as shown in eq 9.14.

$$if\ \chi = \sqrt{\sigma^2 + \varepsilon}\,,\ \beta = \mu\ \ then\ Z^{N(i)} = Z^{(i)} \tag{9.14}$$

Normalizing the input features X can help in learning a neural network. The Batch normalization applies that normalization process not just to the input layer, but the values even deep in some hidden layer in the neural network. So, it will apply this type of normalization to normalize the mean and variance of some of your hidden units' values Z. But, one difference between the training input and these hidden unit values is, hidden unit values be forced to have mean 0 and variance 1. For example, sigmoid activation function, do not want values to always be clustered. Might want them to have a larger variance or have a mean that's different than 0, in order to better take advantage of the non-linearity of the sigmoid function rather than have all your values be in just this linear regime. So, that is why with the parameters χ and β, $Z^{[l](i)}$ values have the range of values. Hidden units have standardized mean and variance, where the mean and variance are controlled by two explicit χ and β. which the learning algorithm can set to range of values. So, what it really does is it normalizes in mean and variance of these hidden unit values, the $Z^{N[i]}$, to have some fixed mean and variance. And that mean and variance could be 0 and 1, or it could be some other value, and it is controlled by these parameters χ and β. For example, consider binary classification problem. In which it has two possible labels, 0 or 1. What if, multiple possible classes? There is a generalization of logistic regression called Softmax regression (Jiang, 2018). The less predictions where you are trying to recognize one of C, rather than just recognize two classes. The output labels $\hat{Y}$ is going to be a $C\ X\ 1$ dimensional vector, because it now has to output C numbers, giving you these C probabilities, the C numbers in the output $\hat{Y}$, they should sum to one. The standard model for getting any network to do this uses what is called a Softmax layer, and the output layer in order to generate these outputs. So, in the final layer of the neural network, compute as usual the linear part of the layers.

9.6 IMPLEMENTATION

To implement more complex models, such as convolutional neural networks or recurring neural networks, is not practical to implement everything yourself from scratch. Fortunately, there are now many good deep learning software frameworks that can help you implement these models (Lin, 2018). So, Deep learning has now matured to that point where it is actually more efficient performing experiments with some of the deep learning frameworks. There are many deep learning frameworks that make it easy for you to implement neural networks, each of these frameworks has a dedicated user and developer community. Each of these frameworks is a credible choice for some subset of applications. There are lot of researchers writing articles in comparing these deep learning frameworks and how good these deep learning frameworks can change. One important criterion is the ease of programming, and that means both developing the neural network and iterating on it as well as deploying it for production. The second important criterion is running speeds, especially training on large datasets, third important is a framework to be truly open. Unfortunately, in the software industry, some companies have a history of open sourcing software but maintaining single corporation control of the software.

The motivation in the experiment is to minimize the cost function as shown in the eq 9.15.

$$\begin{aligned} J(w) &= w^2 + 10w - 25 \\ &= (w-5)^2 \end{aligned} \tag{9.15}$$

with the value $w = 0$, the cost function J will be minimum. Let us find the parameters values $J(w,b)$. So, using Python in Jupyter notebook and to start up TensorFlow (Basha, 2018).

```
import numpy as np
import tensorflow as tf
w = tf.variable (0,dtype = tf.float32)
cost = tf.add (tf.add (w**2, tf.multiply(−10,w)),25)
train = tf.train.GradientDescentOptimizer(0.01).minimize(cost)
init = tf.global_variables_initializer()
session = tf.Session()
print(session.run(w)) ⇒ 0
session.run(train)
print(session.run(w)) ⇒ 0.1

for I in range (1000):
        session.run(train)
print(session.run(w)) ⇒ 4.99999
```

9.7 CONCLUSIONS

Batch normalization makes hyper-parameter search problem much easier, makes neural network much more robust. To implement more complex models, such as convolutional neural networks or recurring neural networks, TensorFlow helps in implementing these models. The cost function considered in the experiment can be trained to get the exact value of hyper-parameter (4.99999) after 1000 iterations. The findings made in the research are: L2 Regularization to be used in reducing variance (0.01), Adam optimization algorithm is used by gradient to update weights. Batch normalization makes hyper-parameter search problem much easier, makes neural network much more robust. In future, we would like to develop huge neural network in training the model quickly to obtain the expected outcome with low variance for different application domains.

KEYWORDS

- **neural network**
- **hyper-parameter tuning**
- **regularization**
- **TensorFlow**
- **bias**
- **variance**

REFERENCES

1. Aljarah, I.; Faris, H.; Mirjalili, S. Optimizing Connection Weights in Neural Networks Using the Whale Optimization Algorithm. *Soft Comput.* **2018,** *22* (1), 1–15.
2. Basha, S. M.; Rajput, D. S.; Vandhan, V. Impact of Gradient Ascent and Boosting Algorithm in Classification. *Int. J. Intell. Eng. Syst.* **2018,** *11* (1), 41–49.
3. Basha, S. M.; Rajput, D. S.; Iyengar, N.; Caytiles, D. R. A Novel Approach to Perform Analysis and Prediction on Breast Cancer Dataset Using R. *Int. J. Grid Distrib. Comput.* **2018,** *11* (2), 41–54.

4. Basha, S. M.; Zhenning, Y.; Rajput, D. S.; Caytiles, R. D.; Iyengar, N. Comparative Study on Performance Analysis of Time Series Predictive Models. *Int. J. Grid Distrib. Comput.* **2017,** *10* (8), 37–48.
5. Chowdhary, C. L.; Acharjya, D. P. Singular Value Decomposition–Principal Component Analysis-Based Object Recognition Approach. *Bio-Insp. Comput. Image Video Process.* **2018,** 323–341.
6. Glaze, C. M.; Filipowicz, A. L.; Kable, J. W.; Balasubramanian, V.; Gold, J. I. A Bias-Variance Trade-Off Governs Individual Differences in Online Learning in an Unpredictable Environment. *Nat. Human Behav.* **2018,** *2* (3), 213.
7. Jiang, M.; Liang, Y.; Feng, X.; Fan, X.; Pei, Z.; Xue, Y.; Guan, R. Text Classification Based on Deep Belief Network and Softmax Regression. *Neural Comput. Appl.* **2018,** *29* (1), 61–70.
8. Jiang, Y. G.; Wu, Z.; Wang, J.; Xue, X.; Chang, S. F. Exploiting Feature and Class Relationships in Video Categorization With Regularized Deep Neural Networks. *IEEE Trans. Pattern Anal. Mach. Intell.* **2018,** *40* (2), 352–364.
9. Jørgensen, T. B.; Wolniakowski, A.; Petersen, H. G.; Debrabant, K.; Krüger, N. Robust Optimization With Applications to Design of Context-Specific Robot Solutions. *Robot. Comput. Integr. Manuf.* **2018,** *53*, 162–177.
10. Kallam, S.; Basha, S. M.; Rajput, D. S.; Patan, R.; Balamurugan, B.; Basha, S. A. K. In *Evaluating the Performance of Deep Learning Techniques on Classification Using Tensor Flow Application. 2018 International Conference on Advances in Computing and Communication Engineering (ICACCE)*, IEEE, June, 2018; pp 331–335.
11. Laurent, T.; Brecht, J. In *Deep Linear Networks with Arbitrary Loss: All Local Minima are Global.* International Conference on Machine Learning, July, 2018; pp 2908–2913.
12. Levine, S.; Pastor, P.; Krizhevsky, A.; Ibarz, J.; Quillen, D. Learning Hand-Eye Coordination for Robotic Grasping With Deep Learning and Large-Scale Data Collection. *Inter. J. Robot. Res.* **2018,** *37* (4–5), 421–436.
13. Li, T. M.; Gharbi, M.; Adams, A.; Durand, F.; Ragan-Kelley, J. Differentiable Programming for Image Processing and Deep Learning in Halide. *ACM Trans. Graph.* **2018,** *37* (4), 139.
14. Li, Y.; Wang, N.; Shi, J.; Hou, X.; Liu, J. Adaptive Batch Normalization for Practical Domain Adaptation. *Pattern Recogn.* **2018,** *80*, 109–117.
15. Lin, Z.; Ota, J. M.; Owens, J. D.; Muyan-Ozcelik, P. Benchmarking Deep Learning Frameworks with FPGA-suitable Models on a Traffic Sign Dataset, 2018.
16. Paul, T.; Chakraborty, A.; Kundu, S. In *Hybrid Shallow and Deep Learned Feature Mixture Model for Arrhythmia Classification.* 2018 Electric Electronics, Computer Science, Biomedical Engineerings' Meeting (EBBT), IEEE, April, 2018; pp 1–4.
17. Peng, X.; Ducru, P.; Liu, S.; Forget, B.; Liang, J.; Smith, K. Converting Point-Wise Nuclear Cross Sections to Pole Representation Using Regularized Vector Fitting. *Comput. Phys. Commun.* **2018,** *224*, 52–62.
18. Pham, H.; Guan, M. Y.; Zoph, B.; Le, Q. V.; Dean, J. Faster Discovery of Neural Architectures by Searching for Paths in a Large Model, 2018.
19. Srinivasan, V.; Sankar, A. R.; Balasubramanian, V. N. In *Adine: An Adaptive Momentum Method for Stochastic Gradient Descent.* Proceedings of the ACM India Joint International Conference on Data Science and Management of Data, ACM, January, 2018; pp 249–256.

20. Tran, K. P.; Castagliola, P.; Celano, G.; Khoo, M. B. Monitoring Compositional Data Using Multivariate Exponentially Weighted Moving Average Scheme. *Qual. Reliab. Eng. Int.* **2018,** *34* (3), 391–402.

CHAPTER 10

Super-Resolution Reconstruction of Infrared Images Adopting Counter Propagation Neural Networks

ANNA SARO VIJENDRAN*

School of Computing, Sri Ramakrishna College of Arts and Science, Coimbatore 641006, India

**Corresponding author. E-mail: saroviji@rediffmail.com*

ABSTRACT

Coupling an infrared image with a visual image for additional spectral information and properly processing the two information streams has the potential to provide valuable information in night and or poor visibility conditions. Due to the restrictions of the infrared capture device, infrared image resolution is generally low when compared with the visible image resolution. Super resolution can generate a High Resolution image from a Low Resolution. In this chapter, a novel method has been proposed to combine the information from the visual and infra-red image of the same scene using Counter Propagation Neural Network (CPN). The CPN extracts high-frequency components from the low resolution infrared image and its corresponding high-resolution visible image. The components are fused to get the up-scaled infra-red image which is taken as the input of the CPN again. The CPN outputs High Resolution infrared image.

10.1 INTRODUCTION

Tracking and recognition of targets based on infrared (IR) images have become an area of growing interest. Thermal IR imagery provides a capability for identification under all lighting conditions including total darkness. Large information is provided by IR image for higher temperature objects and less information for lower temperature objects. Visual image, on the other hand, provides the visual context to the objects. Coupling an IR image with a visual image for additional spectral information and properly processing the two information streams has the potential to provide valuable information in night and/or poor visibility conditions. Information provided by both sensors increases the performance of tracking and the robustness of the surveillance system.

Image resolution is of importance to image processing. Higher image resolution holds more details, which is important to image recognition or image segmentation. Thus, it is desired to obtain a high-resolution (HR) image from its low-resolution (LR) counterpart(s). Due to the restrictions of the IR capture device, IR image resolution is generally low when compared with the visible image resolution. Super resolution (SR) can generate a HR image from a LR image or a sequence of LR images. SR methods can be categorized into two classes, that is, multiframe-based SR and learning-based SR. In the learning-based approach, the HR image can be derived from its corresponding LR image with an image database.

In this chapter, a novel method has been proposed to combine the information from the visual and IR image of the same scene using counter propagation neural network (CPN). The chapter is organized as follows: Section 10.1 provides details on the previous work carried out in this field followed by a brief note on the demerits of the existing methods. Section 10.2 explains the architecture of the CPN network. Section 10.3 describes the algorithm of the proposed reconstruction of IR images using CPN network. Section 10.4 details the modified concept adopting histogram equalization (HE) of image. Performance analysis and results are discussed in Sections 10.5 and 10.6. Sections 10.7 and 10.8 give the conclusion followed by summary of the chapter.

10.2 LITERATURE REVIEW

It is often a vital preprocessing procedure to many computer vision and image processing tasks which are dependent on the acquisition of imaging data via sensors, such as infra-red (IR) and visible. In the night-time environment, the IR camera captures the thermal image of an object. IR image provides rich information for higher temperature objects, but poor information for lower temperature objects. Visual image, on the other hand, provides the visual context to the objects. IR images provide enhanced contrast between human bodies and their environment. IR sensors are routinely used in remote sensing applications. The IR reflectance of objects may be different than for the visible light. With the development of IR sensor technology, the field of application of IR images has widened. IR imaging is most commonly used in the military and security sectors, which use IR imaging to monitor enemies and detect and remove hidden explosives. Recently, the importance of IR images is increasing in autonomous vehicles, which may be a big market in the automobile industry. It is difficult to recognize an object with an only visible image in the nighttime with low illumination. Therefore, IR image-based object recognition is preferred for driver assistance in the nighttime.[11] In spite of the increasing necessity of IR technology, the resolution of an IR image is normally lower than that of the visible image due to the limited nature of IR sensor. Many algorithms for improving the visual quality of IR images have been developed and one among them is the SR method.

SR is a computer vision technique which enhances the resolution of images by reconstructing the HR images from its LR counterpart.[18] SR methods are classified into two types, namely multi-frame super-resolution and single-image SR algorithms. In multi-frame SR algorithms, multiple images capturing the same region of interest is used to estimate the super-resolved output images. This method gives rise to computational complexity due to a large number of input images. Single-image SR algorithms use a single image to reconstruct the HR image.[9] Single-image SR algorithms are broadly classified into two, namely interpolation-based methods[4] and learning-based methods.[7] Interpolation based approaches are the simplest as it determines the unknown pixel by learning an interpolation function. However, these methods will introduce ringing and blurring artifacts. Learning method has become prevalent in recent years due to its numerous advantages. Learning-based algorithm learns a prior knowledge to establish a mapping between LR and HR image patches. The prior is either explicit or implicit subject to its learning strategy. Explicit priors use a mathematical function to learn the relation between LR–HR patch-pairs[19,20] whereas implicit priors learn it from a training dataset. The training dataset can be an external dataset with numerous images randomly collected or from self-examples extracted from the test image itself. The two methods of learning are direct and indirect mapping methods.[2] It depends on the strategy employed for patch recovery.

A review of the existing methods on SR is done to study the merits and demerits. Various approaches for acquiring IR images and corresponding visible images and fusing them to generate the desired IR image have been proposed.[1,10,16] Wei Gan et al.[22] have proposed a method to enhance the IR image by using the correlation of an IR image and its corresponding visible image. Zhao et al. have presented a reconstruction method for super-resolving IR images based on sparse representation.[25,26] Bavirisetti and Dhuli have utilized anisotropic diffusion to decompose the source images into approximation and detail layers and computed final detail and approximation layers with the help of Karhunen–Loeve transform.[1] Dong et al.[5] have applied the convolutional neural network (CNN) technique to SR. Huaizhong Zhanga et al.[12] address quality improvement in IR images for object recognition. This approach is based on image bias correction and deep learning. It is proposed to increase target signature resolution and optimize the baseline quality of inputs for object recognition. Kappeler et al. have proposed a CNN that is trained on both the spatial and the temporal dimensions of videos to enhance their spatial resolution.[13] As an extension of SRCNN, Kim et al.[14] introduced a very deep CNN-based SR with deeper network structure by employing visual geometry group (VGG) network. They are used residual-learning and extremely high learning rates to optimize a very deep network fast. Also, they adopted gradient clipping to ensure the training stability. Ma et al. have proposed a fusion method based on gradient transfer and total variation minimization so that it can keep both the thermal radiation and the appearance information in the source images.[17] The study by Xiaomin et al.[24] combines information from visual images and IR images. K-means clustering is used to divide training patches into multiple clusters. First, the resolution of the IR images is improved by integrating the information from both visible images and IR images. Next, the training patches are divided into several clusters. Multiple dictionaries are learned for each cluster in order to provide a more accurate dictionary. Finally, Soft-assignment based multiple regression reconstructs the HR patch by the dictionaries corresponding to its K-nearest training patch clusters. Choi et al.[3] have proposed a new framework for improving the IR image resolution by using the HR visible image information. In the framework, they adopt learning-based and reconstruction-based SR algorithms to improve the resolution of the IR image. Kiran et al.[15] have proposed a single-image SR algorithm for IR thermal images. This method reconstructs HR image from its LR counterpart without an external database. Han et al.[21] have proposed a CNN-based super-resolution algorithm using the corresponding visible image. Wei et al.[23] have proposed an improved fuzzy clustering and weighted scheme reconstruction framework. HR patches are estimated according to several most accurate dictionary pairs. Finally, these estimated HR patches are integrated together to generate a final HR patch by a weighted scheme. Liu et al.[6] present a Classified Dictionary Learning method to reconstruct HR IR images. The optimal pair of dictionaries is chosen for each image reconstruction. Satisfactory results have been achieved without the increase in computational complexity and time cost.

The sparse representation-based SR method lacks detailed information and gives reconstruction results which are not satisfactory. The dictionary learning methods in SR aim at learning a universal and over-complete dictionary to represent various image structures. A large number of different structural patterns exist in an image, whereas one dictionary is not capable of capturing all of the different structures. The optimization for dictionary learning and image reconstruction requires a highly intensive computation, which restricts the practical application in real-time systems.[26–29] The existing learning-based methods are prone to introduce unexpected details into the resultant HR images and tend to blur fine details. The disadvantages of CNN are its high computational cost, slow to train, and needs a lot of training data.

10.3 COUNTER PROPAGATION NEURAL NETWORK

Counter propagation neural (CPN) networks are very popular because of its simplicity in calculations and strong power of generalization. It is based on the clustering of data, a method by which large sets of data are grouped into clusters of smaller sets of similar data. They are based on unsupervised learning. Further self-organization in CPN networks is a natural-clustering process in which the network performs competitive learning to perceive pattern classes based on data similarity. Hence, learning is fast in the CPN networks.

The CPN network was developed by Robert Hecht-Nielsen. He combined the unsupervised Kohonen layer with a Grossberg layer. This network synthesizes complex classification problems, with a minimum number of processing elements and lesser training time. Counter-posing flow of information through the network structure is the uniqueness. There is only one feedforward path from the input layer to the output layer. It is three-layer architecture. An input vector is applied to the nodes on the first layer. The input nodes and the Kohonen layer and the outputs of the Kohonen layer are connected to the Grossberg layer by connecting weights. The output of the Kohonen layer is "winner take all" and is determined by the net input value. Each node on the second layer calculates its net input value and a competition is held to determine which unit has the largest net-input value. That node is the only node that sends a value to the nodes in the third layer. The output of the Grossberg layer is arrived by scaling the output of the Kohonen layer with the connecting weights to the Grossberg layer. Therefore, for any given input, the output will be generated through two layers of transformation, where the input layer has a fixed number of nodes depending on the input vector size and the output is defined by the application requirements. The number of neurons in the Kohonen layer can be optimized to obtain the desired output accuracy.

The Kohonen layer is trained by a geometrical fitting procedure, where the weight vector is found by comparing the input vector to the original weight vector and applying a training rate coefficient. Once the Kohonen layer has been trained, the Grossberg layer can be trained by applying a training vector with a known classification. The previously trained Kohonen winning node will enable the appropriate weight vector and Grossberg output can be forced to a desired binary output response.

The architecture is shown in Figure 10.1.

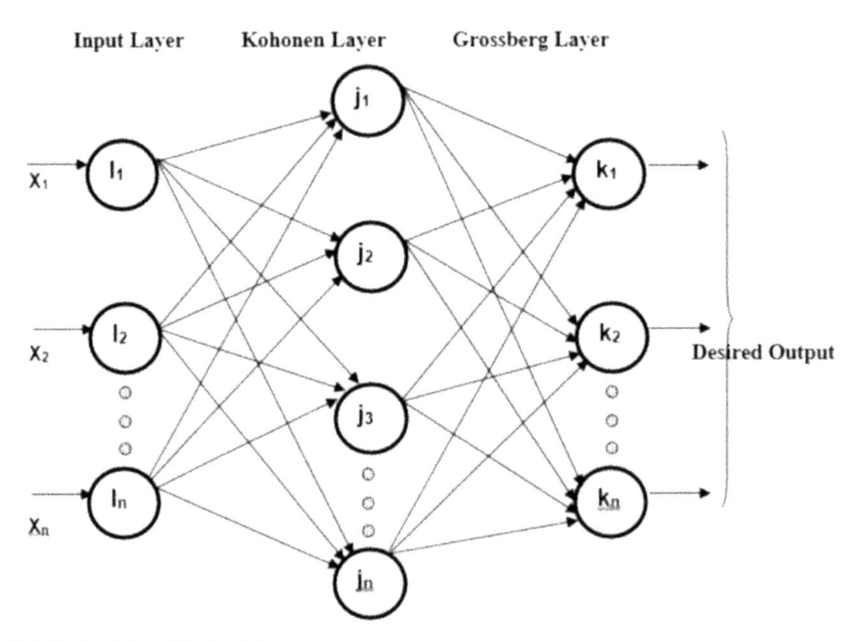

FIGURE 10.1 CPN architecture.

An input vector is applied to the nodes on the first layer. The input nodes of the Kohonen layer are connected to the Kohonen neurons by weights w_{ij} while the Kohonen outputs are connected to the Grossberg layer by the connecting weights v_{jk}. The output of the Kohonen layer is "winner take all" and is determined by the dot product $x_i.w_{ij}$ connections to each node. Each node on the second layer calculates its net input value and competition is held to determine which unit has the largest net-input value. That node is the only node that sends a value to the nodes in the third layer. By scaling the output of the Kohonen layer with the connecting weights to the Grossberg layer, the output of the Grossberg layer is found. Therefore, for any given input, the output will be generated through two layers of transformation, where the input layer has a fixed number of nodes depending on the input vector size and the output is defined by the application requirements. The number of neurons in the Kohonen layer can be optimized to obtain the desired output accuracy.

The Kohonen layer is trained by a geometrical fitting procedure, where the weight vector w_{ij}(new) is found by comparing the input vector x_i to the original weight vector w_{ij}(old) and applying a training rate coefficient α eq 10.1 is iteratively applied through the training process until the trained weights are achieved.

$$w_{ij}(\text{new}) = w_{ij}((\text{old}) + \alpha(x_i - w_{ij}(\text{old})); \; i = 1 \text{ to } n \tag{10.1}$$

Once the Kohonen layer has been trained, the Grossberg layer can be trained by applying a training vector with a known classification. The previously trained Kohonen winning node will enable the appropriate weight vector v_{jk} and the desired output is obtained in the Grossberg Layer. Then the vector weights can be trained by eq 10.2.

$$v_{jk}(\text{new}) = v_{jk}(\text{old}) + \beta(y_k - v_{jk}(\text{old})); \; k = 1 \text{ to } n \tag{10.2}$$

In eq 10.2, β is gradually decremented from 0.1 to 0.0, k is the Kohonen output vector and y_k is the desired output vector. The previous connecting weight is v_{jk}(old) and the new weight is v_{jk}(new).

10.4 SUPER-RESOLUTION RECONSTRUCTION OF INFRARED IMAGES ADOPTING COUNTER PROPAGATION NEURAL NETWORKS

The network is trained using the visible image. The visible image is decomposed into subimages and given as input to the CPN. The network is used to cluster the different subimage vectors into groups, each of which contains similar subimage vectors. The vectors will be grouped together based on their similarities and the number of groups will depend on the required level of enhancement. The quality of the reconstructed image will basically depend upon the total number of allowable groups. For example, if we allow only a few groups then all vectors are bound to fall upon these groups thereby degrading the quality of the image. In order to improve the quality of the image, we have to allow more number of groups. In that case, the space requirement will be increased. So there is a trade-off between total number of groups and total space requirement. The number of neurons "n" in the input layer will be equal to the input vector size X The number of neurons in the Kohonen layer will vary based on the total number of clusters that are allowed. Depending on the distribution and the frequency of gray levels in an image, some of the neurons may not take part in the network simulation. In order to reduce the extra simulation time due to the inactive or dead neurons, the optimum number of neurons in the middle layer is selected by dynamically organizing the neurons in the Kohonen layer. The neurons in the Grossberg layer will also be equal to the size of the input vector. The threshold value is selected such that the trained Kohonen layer will represent the entire range of gray levels of the visible image.

Now, the IR image of the same scene as that of the visible image is given as input to the Trained Neural Network. In IR images, the range of gray values is very low. The network clusters high-frequency components from the low resolution IR image and its corresponding HR visible image. The output from the Grossberg layer will represent an up-scaled infra red image. Now the up-scaled image will be given as the input to the CPN network. Simulation is carried out until the desired HR IR image is obtained.

10.4.1 TRAINING FOR KOHONEN LAYER

An input image is represented in the vector form and normalized.

1. Random weights w_{ij} are assigned between the input layer and the middle layer and v_{jk} between the Kohonen layer and the Grossberg layer.
2. The input vector is applied.
3. Each node on the Kohonen layer calculates its net-input value $\sum x_i.w_{ij}$ and competition are held to determine which unit has the largest net input value. The winner neuron is determined.

4. Calculate α (($x_i - w_{ij}$ (old)) for the winning neuron only and the weights are updated using eq 10.1.
5. Steps 1–5 are repeated until all input vectors have been processed once.
6. Step 5 is repeated until all input vectors have been classified properly. One node will win the competition for all input vectors in a single cluster.

10.4.2 TRAINING FOR GROSSBERG LAYER

Assign a value to the error limit eL.

1. The normalized input vector x_i and the corresponding output vector y_j are applied to the network.
2. The winning Kohonen layer neuron is determined.
3. The weights between the winning Kohonen layer neuron and the neurons in the Grossberg layer are updated using eq 10.2.
4. The error = y_j – desired output is computed. If eL< error, go to step 6. If error > eL, then a neuron in the middle layer is introduced. Go to step 4.
5. Steps 2–4 are repeated until all vectors of all classes map to the desired outputs.

10.5 MODIFIED METHOD ADOPTING HISTOGRAM EQUALIZATION

The conventional approach using CPN network for reconstruction of Infra-red images has the following drawbacks. The number of neurons in the Kohonen layer, which is dynamically organized, varies from image to image. This is due to the type of distribution of the gray-levels in the image. This could be analyzed from the histograms of the images.

For example, it is seen from the histogram of the visible image shown in Figure 10.2 that the number of gray-levels between particular ranges is more than the other levels. In such type of distribution, the number of separable parameters is less. So, if the number of neurons in the Kohonen layer is too many than required, the simulation time takes too long. Also due to the concentration of pixel values to a limited range, some of the neurons in the Kohonen layer may not get fired at all. This leads to a situation where the network is either idle or execute with less efficiency.

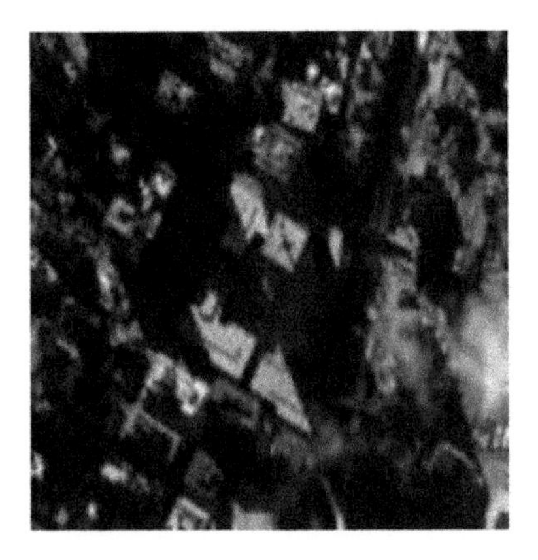

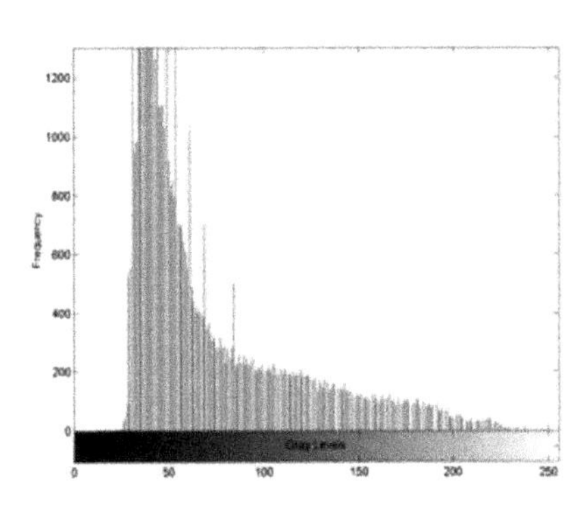

FIGURE 10.2 Visible image and its histogram.

The overall time taken for simulation will be unnecessarily long. In some images, the gray-levels are distributed only for a short range. The problem in the clustering algorithm is that, if the number of clusters is restricted then the representative value of the cluster, which is the average value of the pixels in the cluster, will not be the true representation. While retrieving such information during reconstruction, the original image will get distorted.

To remove these problems, HE is carried out in the visible images. The histogram of the visible image after performing HE is shown in Figure 10.3.

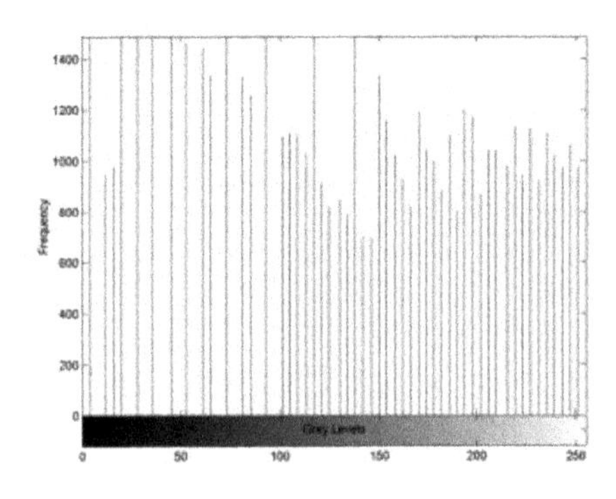

FIGURE 10.3 Histogram equalized visible image and its histogram.

It could be seen from the histogram that equal number of pixels has gray levels in all ranges from 0 to 255. When this image is used in the CPN network, due to the presence of similar pixel values, more number of subimages will be similar. Hence, the formation of clusters will be very fast and thus the training time is drastically reduced resulting in quick convergence of the network. Due to uniform distribution of pixels in various ranges of gray values, we can reduce the number of neurons in the Kohonen layer without distortion to the reconstructed image. Since the visible image now represents the whole range of gray levels, the topology of the input pattern is maintained without much reorganization during this process and hence the reconstruction errors are less and the finer details of the image are preserved.

10.5.1 HISTOGRAM EQUALIZATION

HE is a global contrast enhancement technique for images. HE spreads out and flattens the histogram of image pixels at each gray level value. This method is applied in images in which both the background and foreground are both bright, or both are dark at the same time to get the desired contrast. Consider a discrete grayscale image. Let the number of occurrences of gray level i be n_i. A normalized histogram of the image shows the probability of occurrence of a pixel of level i in the image, and would be given by a collection of probability values for each pixel level, $p(x_i)$ = probability that pixel x has gray level $i = n_i/n$, where n = the total number of pixels in the image. The cumulative density function for this histogram would be given by cdfx $(i) = \Sigma\, p(x_i)$. HE seeks a transformation of input pixel values to output pixel values that will result in the linearization of the cumulative density function across the range of pixel gray levels. That is cdfx $(i) = K\, i$.

10.5.2 PROPOSED CONCEPT

A function f is efficiently learnable if the learning algorithm satisfies the property that for every probability distribution p of the sampling exemplars in X, the algorithm produces a network that will correctly predict the outcome of f with a probability of at least $(1 - \delta)$ for at least a fraction of $(1 - \varepsilon)$ of future exemplars, where δ is the confidence parameter and ε is the error parameter. This means that all future exemplars are assumed to be drawn from the same probability distribution that was used in the learning process.

After a network has successfully learned a set of N training exemplars, the usefulness of the learned network depends on the accuracy of the network's predictions of the output for future exemplars drawn according to the same probability distribution as the training exemplars. The number of training exemplars used in learning is an important factor in the generalization capability of the learned network. A sufficient number of training exemplars are required to describe the function. Good generalization means that the performance observed during training will persist with a high probability, for exemplars not used in training. The distribution of the training exemplars also affects the generalization capability of the learned network.

Let the digital image X, be a subset of an n-dimensional space R^n. Then the pixel values of all n pixels may be represented as a point in R^n. Let f^*: $R^n \rightarrow \{0,1\}$ defines a function, also called concept, in the environment. We are given a set x of training exemplars drawn at random according to some distribution p on $R^n \rightarrow \{0,1\}$. A training exemplar is an input–output pair $(x, f(x)), x \in R^n$.

The goal is to use the set of training exemplars to produce f^*: $R^n \rightarrow \{0,1\}$ that approximates f^* such that with a probability of at least $(1 - \delta)$, $f(x) = f^*(x)$ for at least a fraction $(1 - \varepsilon)$ of future exemplars drawn at random according to the same distribution p. The error probability $e(f)$ of f with respect to p is defined as the probability that $f(x) \neq f^*(x)$, for a random exemplar from the distribution p. The generalized probability $G(f)$ of f is defined as $G(f) = 1 - e(f) = \text{Prob}\ (f(x) = f^*(x))$ for x randomly drawn from X according to the distribution p. Image can be considered as collection of pixels. We define $f^*(x)$ as the cumulative distribution of pixel values = Prob $(X \leq x)$. Now let $y = f^*(x)$ be a function that can be obtained by mapping the random variable X through its own cumulative distribution function. Then,

$$
\begin{aligned}
F_y(y) &= \text{Prob}\ \{f^*(x) \leq y\} \\
&= \text{Prob}\ \{\mathrm{X} \leq \mathrm{f}^*\,\mathrm{x}^{-1}\,(\mathrm{y})\} \\
&= f^*x\,(f^*x^{-1}(y)) \\
&= 0 \text{ for } y < 0 \\
&= y \text{ for } 0 \leq y \leq 1 \\
&= 1 \text{ for } y > 1.
\end{aligned}
$$

This shows that y has a uniform distribution in the interval (0,1). Therefore, Prob $(f(x) = f^*(x))$ is definite for x randomly drawn from X.

Generally, the image will be subdivided into non-overlapping blocks or subimages. Hence, the probability of similar exemplar or subimages is high when the image is mapped by its cumulative distribution function.

Due to this, the convergence of the neural network is not only definite but also very quick. The cumulative distribution function for an image is estimated using the image histogram as follows.

Let $h(i)$ be the histogram of the image formed by computing the number of pixels at gray level i. The pixels will take on the values $i - 0, \ldots, L - 1$ where $L = 256$. The approximation of cumulative distribution function is given by

$$F_x(\mathrm{i}) = \frac{1}{h(L-1)} \sum_{j=0}^{j=1} h(j). \tag{10.3}$$

The normalization term assures that $F_x(L - 1) = 1$. A pixel of X_s is equalized at the position $s \in S$ where S is the set of position in the image.

$$Y_s = F_x(X_s)$$

However, Y_s has a range from 0 to 1 and will not extend over the maximum number of gray levels. To solve, we first compute the minimum and maximum values of Y_s.

$$Y_{max} = \max Y_s, s \in S$$
$$Y_{min} = \min Y_s, s \in S$$

And then we use these values to from Z_s, a renormalized version of $Y_{s.}$

$$Z_s = (L-1)\frac{F_x(X_s) - Y_{\min}}{Y_{\max} - Y_{\min}} \tag{10.4}$$

The transformation from X_s to Z_s results in equal probability associated with every pixel in the image. Accordingly, the modified concept is drawn such that uniform and similar distribution of pixels for any type of image can be obtained when HE is performed in the image. The conclusion drawn from the analysis is that, after applying HE, the similarity among the neighboring pixels is increased. Also due to highly correlated pixels, the number of reference gray-levels required to represent an image is reduced. Hence, when such mapped images are given as input to the artificial neural network, the network converges quickly.

In the modified method, histogram equalized visible images are given as input to the CPN network. HE will reduce the redundancy and increase the information in the visible image. Hence in the CPN network, clustering of high-frequency components from the LR IR image and its corresponding HR visible image will be more efficient and quick. Simulation is carried out until the desired HR IR image is obtained.

10.6 PERFORMANCE EVALUATION OF THE PROPOSED METHOD

In order to evaluate the performance of the proposed method of SR, reconstruction of IR images using CPN network, experimental analysis has been carried out with various images and the results compared with the standard image in terms peak signal to noise ratio (PSNR).

PSNR measure is given by

$$PSNR(dB) = 20 \log \frac{(255/\sqrt{3MN}}{\sum_{i=1}^{M}\sum_{j=1}^{N} A'(i,j) - A(i,j)^{2}}, \tag{10.5}$$

where $A(i,j)$ is the perfect image, $A'(i,j)$ is the reconstructed image.

The proposed method using CPN network has been experimentally tested in various types of images. The first set of experiments was conducted without adopting the HE technique. The results obtained are shown below (Figs. 10.4–10.13).

FIGURE 10.4 (a) Visible image, (b) infrared image, and (c) reconstructed image.

FIGURE 10.5 (a) Visible image, (b) infrared image, and (c) reconstructed image.

FIGURE 10.6 (a) Visible image, (b) infrared image, and (c) reconstructed image.

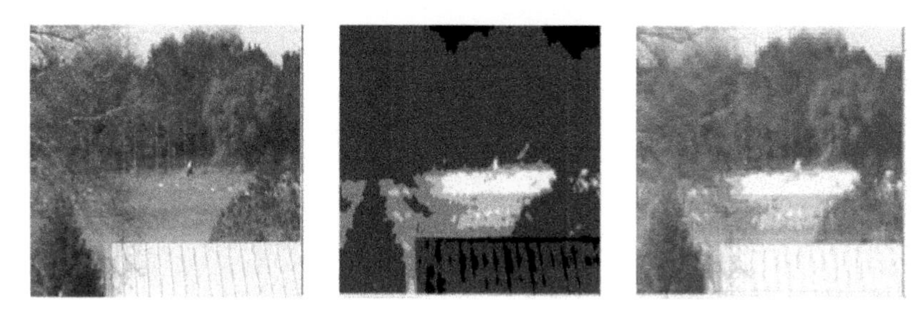

FIGURE 10.7 (a) Visible image, (b) infrared image, and (c) reconstructed image.

FIGURE 10.8 (a) Visible image, (b) infrared image, and (c) reconstructed image.

FIGURE 10.9 (a) Visible image, (b) infrared image, and (c) reconstructed image.

FIGURE 10.10 (a) Visible image, (b) infrared image, and (c) reconstructed image.

FIGURE 10.11 (a) Visible image, (b) infrared image, and (c) reconstructed image.

FIGURE 10.12 (a) Visible image, (b) infrared image, and (c) reconstructed image.

FIGURE 10.13 (a) Visible image, (b) infrared image, and (c) reconstructed image.

The PSNR values of the reconstructed images are tabulated in Table 10.1.

TABLE 10.1 Quality Analyses of Reconstructed Images.

Image	PSNR		
	Visible image	**Infrared image**	**Reconstructed image**
Image 1	36.1672	8.1507	36.8135
Image 2	36.1574	13.5326	38.4943
Image 3	35.0194	14.7449	37.9897
Image 4	35.6898	15.6254	37.8967
Image 5	34.1854	10.5542	36.7528
Image 6	34.1867	12.4562	37.1121
Image 7	34.1922	18.1129	34.9812
Image 8	34.2117	19.5193	34.3758
Image 9	34.1854	20.5181	41.7566
Image 10	34.1901	19.9249	34.7341

The second set of experiments was conducted with the HE visible images. The PSNR values obtained are compared with the values obtained without adopting HE in Table 10.2.

TABLE 10.2 Comparative Analysis of Histogram Equalization Technique.

Image	PSNR	
	Without histogram equalization	**With histogram equalized visible image**
Image 1	36.8135	39.7276
Image 2	38.4943	40.2589
Image 3	37.9897	42.2825
Image 4	37.8967	40.9182
Image 5	36.7528	37.1448
Image 6	37.1121	40.5413
Image 7	34.9812	36.6502
Image 8	34.3758	39.3921
Image 9	41.7566	42.7687
Image 10	34.7341	39.8756

10.7 CONCLUSIONS

CPN network is very popular because of its simplicity in calculations and strong power of generalization. It is based on clustering of data, a method by which large sets of data are grouped into clusters of smaller sets of similar data. CPN is based on unsupervised learning. Further self-organization in CPN is a natural-clustering process in which the network performs competitive learning to perceive pattern classes based on data similarity. Hence, learning is fast. Quality improvement techniques together with faster training procedure have been proposed in this chapter. In the conventional approach, to achieve the desired quality, the number of neurons in the competitive layer is fixed by trial and error procedure. This process not only involves additional computation but also depends on the type of image. So for each and every type of image, the computation has to be done to fix the neurons. To overcome this problem, dynamic organization of the neurons in the competitive layer is proposed. By this approach, the number of neurons can be optimally introduced based on the desired error. In order to further improve the values obtained, HE approach is adopted in the CPN network. Mathematically, a function is efficiently learnable if all future exemplars are assumed to be drawn from the same probability distribution that was used in the learning process. In images, a uniform and equal probability distribution of pixels is obtained when the pixels are mapped by their cumulative distribution function. Accordingly, the new concept was drawn such that uniform and similar distribution of pixels for any type of image can be obtained when HE is carried out in the image. The conclusion drawn from the analysis is that, after applying HE, the similarity among the neighboring pixels is increased. Also due to highly correlated pixels, the number of reference gray-levels required to represent an image is reduced. Hence, when histogram equalized images are given as input to the CPN network, the network converges quickly.

The proposed concept has been implemented and experimentally tested. The experimental results show that the proposed method not only preserves most information of the two images but also enhances the contrast between them. The visual effect of the image is improved efficiently utilizing the proposed method.

10.8 SUMMARY

IR imaging is most commonly used in military and security sectors to monitor enemies and detect and remove hidden explosives. However the resolution of IR image is normally lower than that of visual image due to the limited nature of IR sensor, and blur phenomenon often occurs in the edge area of IR images. Many algorithms for improving the visual quality of IR images have been developed. In this chapter, a CPN network based SR algorithm is proposed to improve the resolution of the IR images. The CPN extracts high-frequency components from the low resolution IR image and its corresponding HR visible image. The components are fused to get the up-scaled IR image which is taken as the input of the CPN again. The CPN outputs High Resolution IR image. Simulation results show that the proposed algorithm outperforms the state of-the-art methods in terms of qualitative as well as quantitative metrics.

KEYWORDS

- **counter propagation neural network**
- **infrared images**
- **super-resolution**
- **visible images**
- **visual image**

REFERENCES

1. Bavirisetti, D. P.; Dhuli, R. Fusion of Infrared and Visible Sensor Images Based on Anisotropic Diffusion and Karhunen-Loeve Transform. *IEEE Sensors J.* **2016,** *16* (1), 203–209.
2. Chang, H.; Yeung, D. Y.; Xiong, Y. Super Resolution Through Neighbor Embedding. *IEEE Comput. Soc. Conf. Comput. Vis. Pattern Recog.* **2004,** *1*, 275–282.
3. Choi, J. S.; Kim, M. Single Image Super-Resolution Using Global Regression Based on Multiple Local Linear Mappings. *IEEE Trans. Image Process.* **2017,** *26* (3), 1300–1314.
4. Dai, S.; Han, M.; Xu, W.; Wu, Y.; Gong, Y. In *Soft Edge Smoothness Prior for Alpha Channel Super- Resolution.* IEEE Conference on Computing Vis. Pattern Classification, Minneapolis, MN, USA, 2007; 1–8.
5. Dong, C.; Loy, C. C.; He, K.; Tang, X. Image Super-Resolution Using Deep Convolutional Networks, *IEEE Trans. Pattern Anal. Mach. Intell.* **2016,** *38* (2), 295–307.
6. Liu, F.; Han, P.; Wang, Y.; Li, X.; Bai, L.; Shao, X. Super-Resolution Reconstruction of Infrared Images Based on Classified Dictionary Learning. *Infrared Phys. Technol.* **2018,** *90*, 146–155.
7. Freeman, W. T.; Jones, T. R.; Pasztor, E. C. Example-based Super-Resolution. *IEEE Comput. Graph. Appl.* **2002,** *22* (2), 56–65.

8. Ghazali, K. H.; Jadin, M. S. In *Detection Improvised Explosive Device (IED) Emplacement Using an Infrared Image*. International Conference on Computer Modelling and Simulation, 2014.
9. Glasner, D.; Bagon, S.; Irani, M. In *Super-resolution From a Single Image*. Proceedings of IEEE International Conference on Computer Vision, IEEE, Kyoto, Japan, 2009.
10. Gyaourova, A.; Bebis, G.; Pavlidis, I. In *Fusion of Infrared and Visible Images for Face Recognition*. European Conference on Computer Vision (ECCV), Berlin Heidelberg, 2004.
11. Han, T. Y.; Song, B. C. In *Night Vision Pedestrian Detection Based on Adaptive Preprocessing Using Near-Infrared Camera*. IEEE International Conference on Consumer Electronics, Asia, 2016.
12. Zhang, H.; Casaseca-de-la-Higuera, P. In *Systematic Infrared Image Quality Improvement Using Deep Learning Based Techniques*. Proceedings of the SPIE, 10008, 2016.
13. Kappeler, A.; Yoo, S.; Dai, Q.; Katsaggelos, A. K. Video Super-Resolution With Convolutional Neural Networks. *IEEE Trans. Comput. Imag.* **2016,** *2* (2), 109–122.
14. Kim, J.; Lee, J. K.; Lee, K. M. In *Accurate Image Super-Resolution Using Very Deep Convolutional Networks*. IEEE Conference on Computer Vision and Pattern Recognition, 2016, Vol. 1, pp 1646–1654.
15. Kiran, Y.; Shrinidhi, V.; Hans, W. J.; Venkateswaran, N. A Single-Image Super-Resolution Algorithm for Infrared Thermal Images. *Int. J. Comput. Sci. Netw. Secur.* **2017,** *17* (10).
16. Li, X.; Qin, S. Y. Efficient Fusion for Infrared and Visible Images Based on Compressive Sensing Principle. *IET Image Process.* **2011,** *52*, 141–147.
17. Ma, J. et al. Infrared and Visible Image Fusion Via Gradient Transfer and Total Variation Minimization. *Inf. Fus.* **2016,** *31*, 100–109.
18. Park, S. C.; Park, M. K.; Kang, M. G. Super-Resolution Image Reconstruction: A Technical Overview. *IEEE Signal Process. Mag.* **2013,** *20* (3), 21–36.
19. Roth, S.; Black, M. J. In *Fields of Experts: A Framework for Learning Image Priors*. Proceedings of IEEE Conference on Computer Vision and Pattern Recognition, USA, 2005.
20. Sun, J.; Shum, H. Y. In *Image Super-resolution Using Gradient Profile Prior*. IEEE Conference on Computer Vision and Pattern Recognition, USA, 2008.
21. Han, T. Y.; Kim, Y. J.; Cheol, B. In *Convolutional Neural Network-Based Infrared Image Super-Resolution Under Low Light Environment*. 25th European Signal Processing Conference (EUSIPCO), 2017.
22. Gan, W.; Ren, C.; Wang, X.; He, X. An Improving Infrared Image Resolution Method via Guided Image Filtering. *J. Softw.* **2014,** *9* (10), 2678–2684.
23. Wei, W.; Yang, X.; Liu, K.; Chen, W.; Zhou, Z. Multi-Sensor Image Super-Resolution With Fuzzy Cluster by Using Multi-Scale and Multi-View Sparse Coding for Infrared Image. *Multimed. Tools Applic.* **2017,** *76* (23), 24871–24902.
24. Yang, X.; Wu, W.; Liu, K.; Zhou, K.; Yan, B. Fast Multi-Sensor Infrared Image Super-Resolution Scheme With Multiple Regression Models. *J. Syst. Arch.* **2016,** *64*, 11–25.
25. Zhao, Y.; Chen, Q.; Sui, X.; Gu, G.. A Novel Infrared Image Super-Resolution Method Based on Sparse Representation. *Infrared Phys. Technol.* **2015,** *71*, 506–513.
26. Chowdhary, C. L.; Ranjan, A.; Jat, D. S. Categorical Database Information-Theoretic Approach of Outlier Detection Model. *Ann. Comput. Sci. Ser.* **2016,** *2016*, 29–36.

CHAPTER 11

High-End Tools and Technologies for Managing Data in the Age of Big Data

NIVEDHITHA* M. and P. M. DURAI RAJ VINCENT

School of Information Technology and Engineering (SITE), Vellore Institute of Technology, Vellore 632014, Tamil Nadu, India

**Corresponding author. E-mail: nivedhithamahendran@gmail.com*

ABSTRACT

In the fast moving world, which would stop if not for the latest technologies, data plays the important role in every aspects. Big data got its name because of the enormous amount of data from various sources, without proper structure. If the data generated makes some sense, it would be greatly useful in various fields for decision making and other prominent areas such as developing Intelligent Systems. Intelligent Systems are capable of perceiving the information and responding to the real world. There are many challenge in doing it, such as analyzing, visualizing and mining the knowledge from the raw data. In this chapter, we have discussed various tools and technologies available to process the unstructured, huge volume of data effectively to gain some insights from them.

11.1 INTRODUCTION

Big data is simply an enormous amount of data that is retrieved from heterogeneous sources (e.g., social media platforms, public, government, e-mail, E-learning platforms, etc.) and is capable of expanding over time. The data available can make a huge impact on the respective fields if it makes some sense, what is the use of having a meaningless data. To make some sense and gain insights for better prediction and decision-making from the boundless noisy data we need proper technologies and tools. An intelligent system is and embedded internet-connected machine which is capable of collecting and analyzing the data also communicate with many other systems. In other words, intelligent systems are machines that are advanced in technology which perceive and respond to the outside world but there are many challenges to develop an intelligent system with all the raw data available. The process of analyzing, visualizing, and mining knowledge from the data is more important than the data itself. In this chapter, we discuss the happening big data technologies which can make noisy, meaningless, a huge amount of data into valuable knowledge, which in turn will help in developing efficient and effective intelligent system.

Attributable to the arrival of much advancement in the internet of things, cloud computing, and many other areas the data are being generated at an unmatched rate from numerous sources. The big data is all about the 4V's (Oussous et al., 2017): Velocity—the rate at which the data is been generated, Volume—the magnitude of the datasets, Variety—the different heterogeneous sources from which the data is been collected, and Value—to generate meaning from the unclear data. In order to acquire and process the data to have a better prediction the traditional databases, which have limited storage and slow data retrieval, it is not a perfect solution.

Nowadays, every field, such as health care, agriculture, retail, weather, generates a huge amount of data that needs proper techniques to be harnessed. Big data technologies offer efficient ways to acquire, store, and process the data effectively that can be used to acquire more knowledge and business profits. The big data technology is classified into two classes: operational and analytical (Sivarajah et al., 2017; Gandomi and Haider, 2015). The operational class deals with the interactive and real-time processing of data and focuses on highly concurrent requests, whereas the analytical class deals with the processing of complex data and focuses on high throughput. Both systems work on the basis of cluster management technique processing the vast amount of data across different cluster nodes.

11.2 PROBLEMS IN MANAGING BIG DATA

The challenges in big data can be classified into data challenge, process challenge, and business challenge (Fig. 11.1) (Gandomi and Haider, 2015; Oussous et al., 2017).

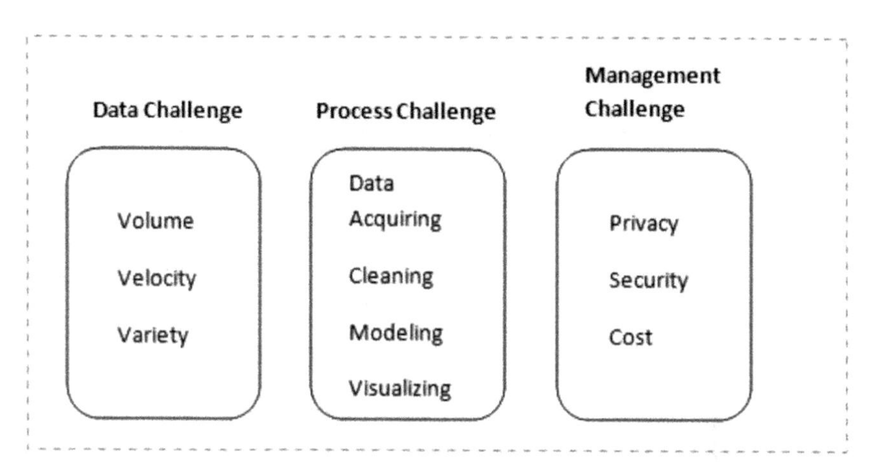

FIGURE 11.1 Hadoop components. Big data challenges.

11.3 DATA CHALLENGES

Data challenges include the V's of big data such as Volume, Velocity, and Variety. We will discuss those V's in this section (Oussous et al., 2017).

- Volume: The definition of volume varies from time to time. Volume literally means the magnitude or size of the data in and out. The data, which is termed as big data now, will not be termed the same in future, over time the definition changes. It is said that Facebook alone stored about 200 Petabyte of photos (one Petabyte = 1024 Terabytes) then think of all the other social media platforms. We have become both the producer and consumers of data, even a mouse click is useful data when processed (Gandomi and Haider, 2015).
- Velocity: Velocity is the time component attached to the data, that is, the data rate at that it is generated and also the time is taken to analyze and gain insight (Oussous et al., 2017). In the present situation where there is a need to process millions of transactions per hour, it is a tedious job to manage. For example, per day on YouTube people upload about hundreds of hours of video.
- Variety: The data collected from random sources may not follow a specific format (video, audio, images, etc.) or structure. The

heterogeneity poses the biggest challenge to process and manage data. About 90% of data generated from different sources are unstructured. For example, the data gathered from social media platforms or data from sensors consists of images, audios, and videos that are an unstructured form of data and are challenging.

11.3.1 PROCESS CHALLENGES

The challenges that are faced while processing the data are termed as process challenges. Processing a large amount of unstructured noisy data is really challenging. The first challenge would be to acquire data from various sources and storing them in a data warehouse (Oussous et al., 2017). Then extracting and cleaning the enormous pool of data to get what is required because in critical areas, such as healthcare one wrong decision can cost patient's life. Once the data has been gathered, cleaned, and extracted it must be modeled and analyzed to gain proper insight. The final challenge, in processing a data, is to visualize the data modeling results and interpret the findings for making some sense from the unclear data.

11.3.2 MANAGEMENT CHALLENGES

Management challenges include (Gandomi and Haider, 2015; Oussous et al., 2017) Privacy: Many companies have invested enormously into big data and have many confidential pieces of information, including passwords, bank account details and managing them is tedious, Security: If there is no proper security control is placed then big data would not be accepted globally. The security challenges, such as availability, integrity, and confidentiality must be handled properly. Cost: To handle the big data cost-effectively, is one of the major challenges as the data is increasing exponentially.

11.3.3 BIG DATA TECHNOLOGIES

To handle all the above-mentioned challenges without proper technologies and tools is a nightmare. Big data offers near-perfect solutions in the form of the following technologies. In this section, we are going to discuss the following happening technologies in big data.

- Hadoop
- Data Storage Layer (HDFS Hbase)
- Data Processing Layer (MapReduce, YARN)+
- Data Querying Layer (PIG)
- Data Access Layer (Flume)
- Data Streaming Layer (Storm, Spark)
- Data Analytics (Mahout)

11.3.4 HADOOP

Hadoop, which is highly archive distributed object-oriented programming, was initially developed for managing a distributed search engine program by Apache Foundation in the year 2005. Hadoop is an open-source JAVA-based programming framework to manage distributed storage and process considerably huge datasets on commodity hardware. It is based on the Master-Slave node architecture.

Hadoop mainly has three components (Fig. 11.2) (White, 2012):

1. HDFS
2. Map Reduce
3. YARN

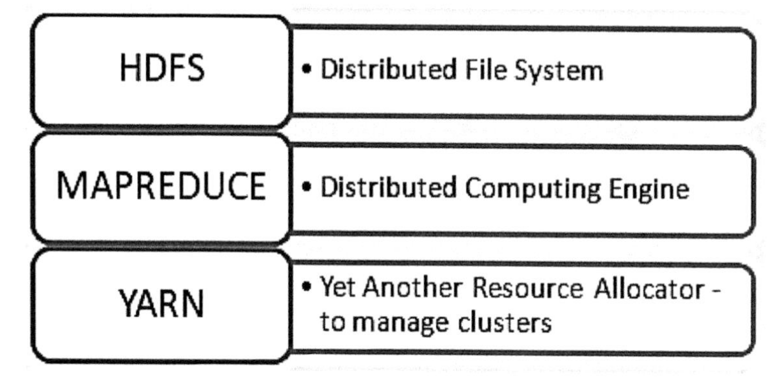

FIGURE 11.2 Hadoop components.

11.4 DATA STORAGE LAYER

11.4.1 HDFS

Hadoop distributed file system is developed based on the policy of write-once and read-many-times where files are stored as clusters across different servers in a redundant fashion so at the time of failure can be recovered. The files are partitioned into different blocks of size 128 MB and are spread across different servers (White, 2012; Saraladevi et al., 2015). HDFS reduces the risk of data loss as it replicates the block on a different machine with a default replication factor 3, so, even if one block is lost the data can be retrieved from other blocks.

The HDFS architecture consists of NameNode and DataNode (Fig. 11.3) (White, 2012; Saraladevi et al., 2015). The NameNode consists of the namespaces and metadata for faster access. It handles all the permissions, such as when the client wants to read/write data from a file NameNode needs to be contacted it also handles requests, such as the creation of files, file storage locations. The DataNode has the actual file data in the form of blocks. The DataNode sends the NameNode two messages, one is Heartbeat to inform about its functioning and another one is BlockReport that has the complete list data blocks.

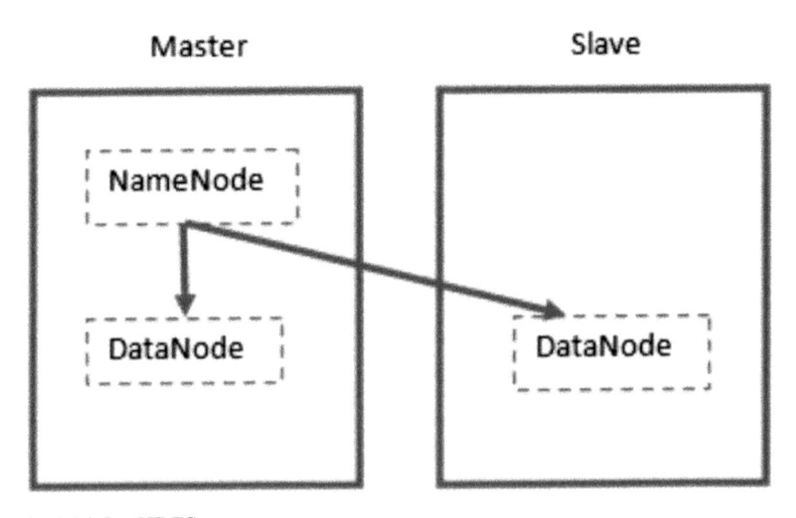

FIGURE 11.3 HDFS components.

11.4.2 APACHE HBASE

Hbase is a column-oriented, highly-scalable; fault-tolerant NoSQL, open-source database built on top of HDFS. Hbase is mainly developed for real-time read/write of large datasets. The data in Hbase are organized as tables with rows and columns. Each row consists of a key to sort and an arbitrary number of columns (Bhupathiraju and Ravuri, 2014; Vora, 2011). The cells in the table are versioned and each version has a timestamp attached to it so that every column can have several versions for the same row key. Columns in every row are grouped by a common column family. The tables are partitioned horizontally into regions.

Hbase cluster architecture consists of three components. They are as follows (George, 2011; Vora, 2011):

1. Hbase master,
2. Zookeeper, and
3. Region servers.

- Hbase master: It is a lightweight process that does not require too many hardware resources. It coordinates with the Hbase cluster in administrative operations, such as creating, updating, and deleting the tables. If a client wants to change a particular schema or meta-data it must be done with the permission of Hbase. In case of large clusters, many Hbase components are created in order to avoid a single point of failure. Only one Hbase master will be active in the server rest all will be synced and when there is a failure in one Hbase master the next one will be chosen by Zookeeper ensemble.
- Zookeeper: It is a lightweight centralized service for managing configuration information, group services, naming, and distributed synchronization. The main purpose of the zookeeper is to assign regions (table partitions) to region servers which are functioning. If a client wants to access either Hbase Master or Region server it must communicate with the zookeeper first as the Hbase Master or region servers are registered with zookeeper. Zookeeper has the record of all the region server activities, so when Hbase master needs information about a particular region server it must contact the zookeeper. Zookeeper also tracks the failure in the region servers and sends an alert message also starts repairing the nodes.

- Region servers: They have a collection of regions which are horizontal partitions of a table. The region servers are called as the worker nodes and handle the read, write, delete requests from the clients. The components of region server are Block cache (read Cache) where the read data is stored, MemStore (Write Cache) where new data which is yet to be written to the disk is stored, Write Ahead Log (WAL) is a file that store new data which is not meant to be in permanent storage, HFile stores rows in the form of sorted keys in the disk which is the actual storage file.

11.5 LIMITATIONS

- No cross data or join operations can be done. The join operation can be implemented only by MapReduce that is time-consuming.
- It will take a while to replace the failed Hbase master with another one,
- As Hbase does not support SQL it is tough to query.
- Migrating data from an external relational database system to Hbase will require a new design.

11.6 DATA PROCESSING LAYER

11.6.1 MAPREDURE

HDFS is for storage, whereas MapReduce is to process huge datasets in parallel across distributed computing environment. The MapReduce consists of two components (Fig. 11.4) (Uzunkaya et al., 2015; Ghadiri et al., 2016):

- JobTracker
- TaskTracker

The JobTracker schedules the tasks so that they can be run by TaskTracker. It also assigns and monitors tasks on all the nodes, when there is a failure in one particular node the JobTracker relaunch the task. The TaskTracker runs the tasks across different nodes and reports them to the JobTracker.

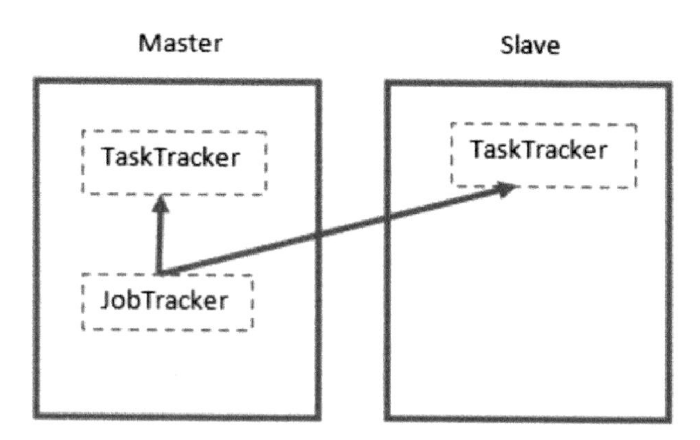

FIGURE 11.4 MapReduce components.

The MapReduce engine has two phases (Nghiem and Figueira, 2016):

- Map phase
- Reduce phase

The map function is to prepare and set the data so that reduce function can work on it. It divides the data into chunks that are later combined in the reducer function. It takes a key-value pair as input and sends an intermediate key-value pair (Ghadiri et al., 2016; Nghiem and Figueira, 2016) as output that is given as input to the reduce function. The reduce function has three phases: sort, shuffle, and reduce. The sort and shuffle phase happens simultaneously wherein shuffle the data are transferred from the mapper to the reducer and in sort phase, the keys from mapper functions are grouped to input the reducer function and in reduce phase the computation of data takes place (Nghiem and Figueira, 2016).

11.6.2 YARN

YARN—yet another resource negotiator—was initially developed to improve the computational abilities of MapReduce engine. It acts as a resource manager and offers APIs for requesting and working with the clusters. The user may not know the resource management details which are hidden. YARN splits the jobs of TaskTracker and JobTracker from the MapReduce into four components (White, 2012; Khan et al., 2015):

1. The resource manager
2. Application master

3. The node manager
4. The container

In step 1, the client creates an end-user application which sends a request to the resource manager that manages the hardware resources and allocates user jobs to those resources. Once the request is valid the step 2 consists of the resource manager contacting the application master to allocate a container to each node that is a resource allocation unit. In step 3 the application master negotiates with the resource manager for deciding the number of resources to be allocated for that particular request. Then in the final step, the resource manager allocates the required resources to process the request from the client application. The node manager is to update the resource manager with the cores that are processing the request and also the memory available in the form of heartbeats. The heartbeat contains the status of the request from the client application (Fig. 11.5) (White, 2012; Khan et al., 2015).

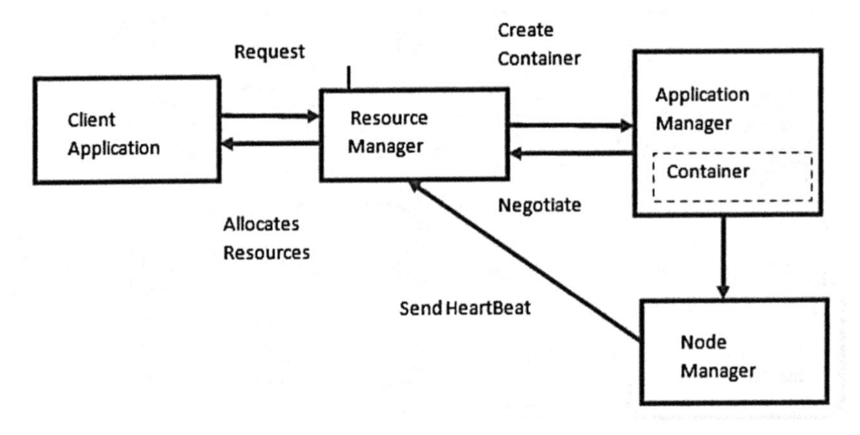

FIGURE 11.5 YARN workflow.

11.6.3 LIMITATIONS OF HDFS AND MAPREDUCE

- Poor availability. HDFS is, of course, offers high reliability and fault-tolerance but when there is a failure in the NameNode the snapshot from SecondaryNameNode replaces the failed node which is done manually and takes a lot of time which makes the system's availability poor.

- Security. As many companies consider Hadoop nowadays to store their critical information including passwords, credit card numbers, and so on. Hadoop, in turn, must provide strong security.
- Hadoop is not suitable for numbers of small datasets
- The shuffle phase in MapReduce is the time-consuming phase that may even overload the network and results in failure.

11.7 DATA QUERYING LAYER

11.7.1 APACHE PIG

Apache PIG is a platform that works on top of HDFS to provide an abstraction over MapReduce in order to reduce its complexity (Vaddeman, 2016). The data manipulation operations can be performed with ease using PIG on Hadoop. PIG has a simple language to analyze data called as PIG Latin that has rich sets of operations and data types. The developers write a script using PIG Latin which is then converted to a series of MapReduce jobs internally which makes the developer's job easy. Apache PIG handles all kinds of data and also automatically optimizes the data before its execution.

The Pig Data model consists of (Jain and Bhatnagar, 2016): (1) Atom any value in the pig Latin script is considered as an atom irrespective of the data types that are by default stored as a string then used in many other forms. (2) A tuple is an ordered set of fields that can be of any type. (3) The bag is an unordered collection of tuples. (4) Relation—a bag of tuples, and (5) Map—set of key-value pairs where the key is unique and value can be of any type.

The architecture of PIG consists of the following components (Vaddeman, 2016; Gates and Dai, 2016):

1. Parser
2. Optimizer
3. Compiler
4. Execution engine

- Parser: The parser parses through the Pig Latin scripts and checks for syntax errors and type checking and other checks. The output is sent to the optimizer, which is in the form of directed acyclic graph, that has Pig Latin statements and logical operators.

- Optimizer: The optimizer performs logical optimizations, such as projection, and pushes down on the received directed acyclic graph and forwards it to the compiler,
- Compiler: The compiler converts the logical optimizations received from the optimizer into a series of MapReduce jobs and sends it to the execution engine.
- Execution Engine: The series of MapReduce tasks are sorted and submitted to the Hadoop which are executed by the execution engine and outputs the desired result.

11.7.2 LIMITATIONS

- Pig scripts always produce the same exact error even if there are some other problems like syntax or logical.
- It can be used only on semistructured data.
- As commands are not executed unless they are stored as an intermediate the debugging process takes long.

11.8 DATA ACCESS LAYER

11.8.1 APACHE FLUME

Flume became Flume NG (New Generation) when it was donated to the Apache community by Cloudera. Flume is a distributed service that offers a real-time collection of data and stores the collected data temporarily and delivers it to the target when needed (Oussous et al., 2017). The main purpose of the Flume is to collect and transfer the streaming data from several servers to HDFS.

The Flume architecture consists of four components that run inside a daemon called agent (Fig. 11.6) (Birjali et al., 2017). They are as follows:

1. Source
2. Channels
3. Sink
4. Agent

- Source: Flume gathers messages from different sources such as spooling directory source, Syslog file source, Exec source, and many more (Birjali et al., 2017). A source can be able to write on multiple channels using a channel selector. A source is simply an input to the channels.
- Channels: Channels are the temporary storing area for the messages collected from different sources. It is the intermediate between the source and the sink. There are two types of channels memory-backed/nondurable channel and local filesystem-backed/ nondurable channel (Hoffman, 2013). The memory-backed channels are faster when compared to the filesystem-backed channels as memory retrieval is faster than the disk retrieval. The disadvantage with this is that when there is an agent failure, such as power problem, hardware crash, or flume restart there would be a data loss. Whereas the filesystem-backed channels the data is stored in the local file system that makes retrieval of data slower but durable. These channels can withstand agent failures and considerably less amount of data loss. Finally, the Channel outputs the message to the sink.
- Sink: Sink receives the data coming from the channels and writes them to the target, such as HDFS, MongoDB, and Cassandra, etc. A sink is simply output from to be written on destination
- Agent: An agent is a daemon tool that encapsulates the source, channels, and sink. A Flume agent can have more than one source, channel, and sink.

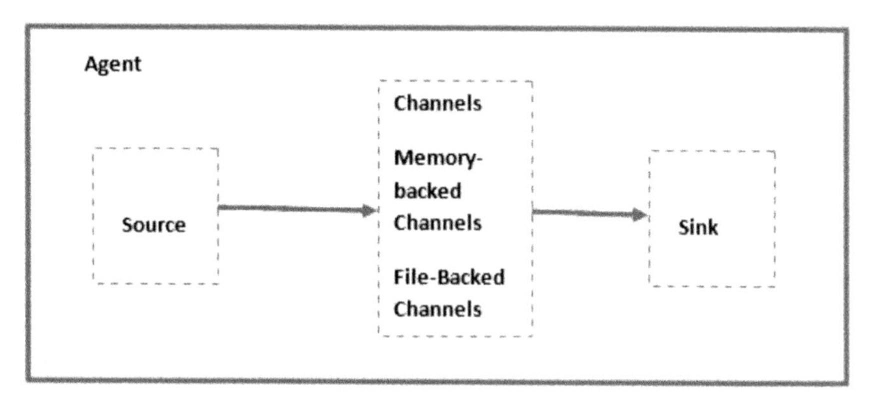

FIGURE 11.6 Flume workflow.

11.8.2 LIMITATIONS

- The message that reaches the destination is not guaranteed to be unique. There is a possibility of duplicate messages being delivered.
- The throughput of flume depends on the channel so scalability and reliability are often placed under the lens.

11.9 DATA STREAMING LAYER

11.9.1 APACHE SPARK

Apache Spark is a fast and general-purpose cluster computing platform to extend the MapReduce framework to process interactive queries and stream processing. The important aspect of processing an enormous amount of data is speed, as Spark offers in-memory computation where the data is stored in the main random access memory. The in-memory computing offers a faster response and has the ability to handle complex computations (Oussous et al., 2017). As Spark uses this computing technique it is comparatively faster than the MapReduce for processing complex data. Spark runs upon Hadoop clusters and can easily be integrated with any of the big data tools that makes combining a wide range of workloads easy and fast.

Spark consists of six components to manage memory, fault-tolerance, schedule tasks, interactive processing, and so on (Karau et al., 2015). They are as follows:

1. Spark core
2. Spark SQL
3. Spark streaming
4. MLlib
5. GraphX
6. Cluster managers

- Spark Core: The spark data model is build based on the resilient distributed dataset (RDD) abstraction that consists of read-only objects distributed across different machines. Spark core offers APIs to access and manipulate those RDD read-only objects. Spark core also consists of components to schedule tasks, manage memory, fault recovery, and so on.
- Spark SQL: Combines RDD and the relational database to form spark SQL that makes querying the external dataset and complex

analysis easy and efficient. Spark SQL allows the programmer to analyze the query from external sources and also from existing data in RDDs. It supports many sources such as HIVE, JSON, and many more for efficient parallel processing of data. It can be integrated with Python, Scala, and java all in one application that makes it one among the strongest tools in data warehouse.

- Spark Streaming: This component of spark enables live streaming of data, such as consistent live status updates from the users in social media platform. It offers APIs for users to write batch jobs in Java, Scala, which are then streamed. It also makes the integration of batch jobs and interactive queries easier. Spark streaming, like Spark core, has components for fault-tolerant, memory management, and job scheduling.
- MLlib: MLlib is a framework built on top of spark for supporting machine learning algorithms such as classification, regression, clustering, and many more. Like Apache Mahout, MLlib also supports machine learning techniques the difference is mahout does not support the regression models and MLlib does.
- GraphX: This component is to help in manipulating the graphs in the form of graph-parallel computing. GraphX extends the RDD API so that user can create directed graphs with vertex and edges. It also provides operators, such as subgraph and mapVertices, for manipulating graphs and library of graph algorithms, such as PageRank and triangle counting.
- Cluster Managers: Spark supports many cluster managers, such as Hadoop YARN, Mesos clusters, etc., when there is no cluster manager in the system where Spark is installed the spark uses the in-built default cluster manager called Standalone Scheduler where the main purpose is to scale up the resources.

11.9.2 LIMITATIONS

- Apache Spark does not support real-time processing. The data collected through live streams are processed using batch processed in batches using map and reduce functions.
- When Spark is used with Hadoop there is a problem with handling small files. As HDFS cannot handle large number of small files
- Spark does not have file managing system of its own, it depends on other platforms, such as Hadoop and Cloudera, for managing files.

- Expensive.

11.10 APACHE STORM

Apache Storm is a distributed system for handling real-time data, whereas Hadoop is only for batch processing. The storm was developed for fault-tolerant and horizontal scaling of the vast amount of data. Apache storm manages the cluster state in the distributed environment using Apache Zookeeper. The storm can continue the process even at the time of failure by adding the resources in a linear manner.

The components of Apache Storm are as follows (Fig. 11.7) (Evans, 2015; Batyuk and Voityshyn, 2016):

1. Tuples, which is an ordered list of any type of comma separated values, are passed to the storm cluster.
2. Stream that in unordered collection of tuples.
3. The spout is a source of the stream that can accept data from external sources, such as twitter APIs, and puts that into topology or one can write a spout to read data. A spout can be of two forms reliable that can replay a tuple in case of failure the other one is unreliable that forgets as soon as emitted. ISpout is a common interface that implements Spout.
4. Bolt all the processing of data is done in the bolts section like filtering, joins, function, and much more. Spout passes the data on to Bolt and it processes them and emits to one or more bolts. IBolt is a common interface that implements the Bolts.
5. Topology is the package of logics that is needed for real-time applications. It consists of all the spouts and bolts in the form of a graph connected with stream groupings.

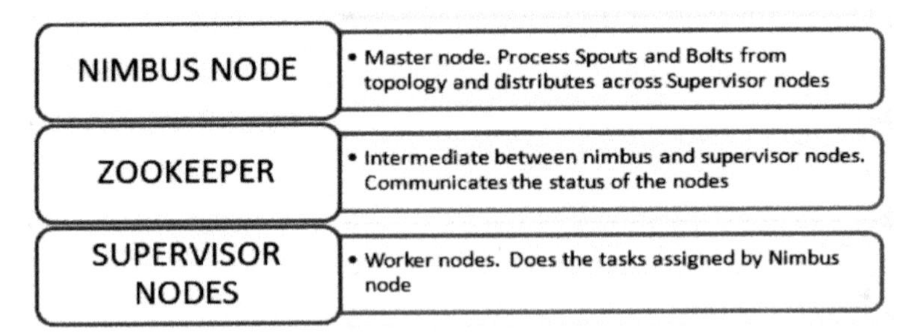

FIGURE 11.7 Storm components.

The architecture of Storm consists of Nimbus (Master Node), Zookeeper and supervisors (Worker Node) (Batyuk and Voityshyn, 2016).

The Nimbus, which is a master node, will receives the storm topology that has spouts and bolts then it gathers all the tasks and distributes them to the supervisor nodes which are Worker nodes. Then the supervisor nodes send a heartbeat to the nimbus node as an acknowledgment of their status when there is no heartbeat the nimbus assumes that the supervisor node is dead and assigns that task to some other node. In case the nimbus is dead the supervisor nodes will work even then with the already assigned tasks and in the meantime, nimbus will be restarted. When all the topologies are completed the nimbus and supervisor will wait for new topology and tasks. As nimbus is stateless it cannot monitor the working nodes without the help of zookeeper. The Zookeeper acts as an intermediate between the nimbus and supervisor nodes and communicates the state of working nodes

11.10.1 LIMITATIONS

- Updating a topology that is running cannot be done. To update a running topology it must be killed then updated.
- Apache storm is stateless it depends on Zookeeper to save the states of its working nodes.

11.11 DATA ANALYTICS LAYER

11.11.1 APACHE MAHOUT

Apache Mahout is a scalable open source project mainly developed for implementing machine learning techniques. Machine learning is a technique where the algorithms are trained based on past experiences and tested to predict a near-perfect future. Mahout runs along with Hadoop to manage huge volumes of data the algorithms are written on top of Hadoop. Mahout offers ready-to-use algorithms for data mining tasks. Apache Mahout supports mainly collaborative filtering to mine user behaviors and recommendation systems, clustering where similar items are grouped together, classification to categorize the data into the specific class (Giacomelli, 2013; Meng et al., 2016).

The Machine Learning algorithms are of three forms (Malhotra, 2015):

1. Supervised learning
2. Unsupervised learning
3. Reinforcement learning

- Supervised learning: In supervised learning, the input and output results are known already. The algorithm receives the set of inputs and known set of outputs then it is made to learn with past data. This model is most useful when there is a historic data and future events need to predict based on that data. Supervised learning is of two forms—classification that has continuous data and regression that has discrete. Apache mahout supports only classification machine learning algorithms.
- Unsupervised learning: In unsupervised learning, the input is known but the output is inferred from the input given. The unsupervised learning techniques work well for transactional data. Unsupervised techniques are mainly based on clustering.
- Reinforcement learning: This is mainly for gaming and robotics where there is no explicit force to instruct the algorithm of what to do and what not. The algorithm itself will learn based on the environment and earn reward points. This is done by trial and error methods. The main components of reinforced learning are learner, environment, and actions.

11.11.2 LIMITATIONS

- Poor scalability.
- When there is a retraining of the algorithm the whole recommender needs to be transferred to the node that causes network traffic and also time-consuming.

11.12 CONCLUSIONS

Intelligent systems in common terms is a machine, which is embedded with many other technologies and also connected to the internet, capable of analyzing data and communicating with other such systems in the network that also leads to a heavy amount of data generation. Intelligent systems are in trend in many industries, such as transportation, logistics, security, marketing, hospitals, manufacturing and much more. They are more complex and incorporates more technologies, such as cybersecurity, artificial intelligence, deep learning, embedded systems, networking,

TABLE 11.1 Comparison of big data tools.

Layer	Tools	Functionality	Limitations
Data storage layer	HDFS	• Write-once and read-many-times • Stored as clusters across different servers in a redundant fashion • Recovery is easily done in case of failure	Poor availability at the time of failure
	Hbase	• Column-oriented, highly scalable, fault-tolerant NOSQL • Real-time read/write of large datasets	• Tough to query • Migrating data from external RDMS is tedious
Data processing layer	MapReduce	• Process large datasets in parallel across a distributed environment.	• Not suitable for small datasets. • Shuffle phase is time-consuming may lead to network overhead.
	YARN	• Resource manager • Offers API's for working with clusters.	• Uncertainty. YARN does not ensure when the job will be completed.
Data Querying Layer	PIG	• Provides abstraction over MapReduce to reduce complexity • Analyze data using a simple language PIG LATIN	• Can be used only on semi-structured data • Debugging takes more time.
Data access layer	FLUME	• Collect and transfer the streaming data from several servers to HDFS	• Possibilities of duplicate message delivery
Data streaming layer	Spark	• Interactive queries and stream processing • In-memory computation	• Doesn't support real-time processing • Expensive • Depends on other platforms to manage files

TABLE 11.1 *(Continued)*

Layer	Tools	Functionality	Limitations
	Storm	• Handles real-time data • Fault-tolerant, horizontal scaling of vast amount of data	• Updating a running topology cannot be done • Stateless, depends on zookeeper to save the states.
Data analytics layer	MAHOUT	• Implements machine learning algorithms	• Poor scalability • Network overhead at times, time-consuming in retraining of an algorithm.

distributed storage, and many. Due to the availability of many sources for data generation, such as streaming data, social media platforms, click-stream data, and many advanced fields, such as internet of things, cloud computing, there is a huge amount of data being generated every second online and offline. Managing such volume of data and deriving knowledge from them is really a tedious job at present situation without effective tools and technologies. To derive proper knowledge and employ them in making a better decision or discovering pattern can be done effectively.

An intelligent system is something that should be capable of learning from experience, security, connectivity, and able to adapt to the current data and capable of remote monitoring management. With all those noisy unstructured data it is very difficult to develop an intelligent system. There are many challenges in developing an intelligent system such as uncertainty in the data collected from sensors or many other sources, a dynamic world which requires decisions to be made at a faster rate, time-consuming computations, and mapping. Such a huge amount of structured and unstructured data is termed as big data and there are many big data technologies available in the market to handle such data. The big data tools help in managing data that are varied, diverse, dynamic, and massive. They help in making the operations faster and more accurate intelligent decisions. The big data technologies are usually put under two classes one is operational and the other is analytical where the former focuses on concurrent requests with low latency and the later focuses on high throughput. The above-discussed tools and technologies come under operational and analytical classes make a huge difference than the traditional technologies in handling the challenges in developing intelligent systems and other systems for that matter.

KEYWORDS

- **intelligent systems big data**
- **hadoop ecosystem mapreduce**
- **pig**
- **hive**
- **apache storm**

REFERENCES

1. Batyuk, A.; Voityshyn, V. *Apache Storm Based on Topology for Real-time Processing of Streaming Data from Social Networks*. In Data Stream Mining & Processing (DSMP), IEEE First International Conference on IEEE, August, 2016, pp 345–349.
2. Bhupathiraju, V.; Ravuri, R. P. *The Dawn of Big data-HBase*. In IT in Business, Industry and Government (CSIBIG), 2014 Conference on IEEE, March, 2014, pp 1–4.
3. Birjali, M.; Beni-Hssane, A.; Erritali, M. Analyzing Social Media through Big Data using InfoSphere BigInsights and Apache Flume. *Procedia Com. Sci.* **2017,** *113*, 280–285.
4. Evans, R. *Apache Storm, A Hands on Tutorial*. In Cloud Engineering (IC2E), 2015 IEEE International Conference on IEEE, March, 2015, pp 2–2.
5. Gandomi, A.; Haider, M. Beyond the Hype: Big Data Concepts, Methods, and Analytics. *Int. J. Inform. Manage.* **2015,** *35* (2), 137–144.
6. Gates, A.; Dai, D. *Programming Pig: Dataflow Scripting with Hadoop;* O'Reilly Media, Inc., 2016.
7. George, L. *HBase: The Definitive Guide: Random Access to your Planet-size Data;* O'Reilly Media, Inc.,2011.
8. Ghadiri, N.; Ghaffari, M.; Nikbakht, M. A. *BigFCM: Fast, Precise and Scalable FCM on Hadoop*. ArXiv preprint arXiv:1605.03047.
9. Giacomelli, P. *Apache Mahout Cookbook;* Packt Publishing Ltd.: Birmingham, UK
10. Hoffman, S. *Apache Flume : Distributed Log Collection for Hadoop;* Packt Publishing Ltd.: Birmingham, UK, 2013.
11. Jach, T.; Magiera, E.; Froelich, W. Application of HADOOP to Store and Process Big Data Gathered from An Urban Water Distribution System. *Procedia Eng.* **2015,** *119*, 1375–1380.
12. Jain, A.; Nalya, A. *Learning Storm*. Packt Publishing, 2014.
13. Jain, A.; Bhatnagar, V. Crime Data Analysis Using Pig with Hadoop. *Procedia Com. Sci.* **2016,** *78*, 571–578.
14. Karau, H.; Konwinski, A.; Wendell, P.; Zaharia, M. *Learning Spark: Lightning-fast Big Data Analysis;* "O'Reilly Media, Inc.: Sebastopol, California, 2015.
15. Malhotra, R. A Systematic Review of Machine Learning Techniques for Software Fault Prediction. *Appl. Soft Com.* **2015,** *27*, 504–518.
16. Meng, X.; Bradley, J.; Yavuz, B.; Sparks, E.; Venkataraman, S.; Liu, D.; Xin, D. Mllib: Machine Learning in Apache Spark. *J. Mach. Learning Res.* **2016,** *17* (1), 1235–1241.
17. Nghiem, P. P.; Figueira, S. M. Towards Efficient Resource Provisioning in MapReduce. *J. Parallel Distr. Com.* **2016,** *95*, 29–41.
18. Oussous, A.; Benjelloun, F. Z.; Lahcen, A. A.; Belfkih, S. Big Data Technologies: A Survey. *J. King Saud University-Com. Inform. Sci.* **2018,** *30* (4), 431–448
19. Saraladevi, B.; Pazhaniraja, N.; Paul, P. V.; Basha, M. S.; Dhavachelvan, P. Big Data and Hadoop-A Study in Security Perspective. *Procedia Com. Sci.* **2015,** *50*, 596–601.
20. Sivarajah, U.; Kamal, M. M.; Irani, Z.; Weerakkody, V. Critical Analysis of Big Data Challenges and Analytical Methods. *J. Business Res.* **2017,** *70*, 263–286.
21. Uzunkaya, C.; Ensari, T.; Kavurucu, Y. Hadoop ecosystem and its analysis on tweets. *Procedia-Social Behav. Sci.* **2015,** *195*; 1890–1897.
22. Vaddeman, V. *Beginning Apache PIG;* Apress: Berkeley, CA, 2016

23. Vora, M. N. *Hadoop-HBase for Large-scale Data.* In Computer science and network technology (ICCSNT), 2011 international conference on IEEE, December 2011, Vol. 1, pp 601–505.
24. Weets, J. F.; Kakhani, M. K.; Kumar, A. *Limitations and Challenges of HDFS and MapReduce.* In Green Computing and Internet of Things (ICGCIoT), 2015 International Conference on IEEE, October, 2015, pp 545–549.
25. White, T. *Hadoop: The Definitive Guide;* O'Reilly Media, Inc., Sebastopol, California, USA, 2012.
26. Zafar, H.; Khan, F. A.; Carpenter, B.; Shafi, A.; Malik, A. W. MPJ Express Meets YARN: Towards Java HPC on Hadoop Systems. *Procedia Com. Sci.* **2015,** *51*, 2678–2682.

CHAPTER 12

An AI-Based Chatbot Using Deep Learning

M. SENTHILKUMAR and CHIRANJI LAL CHOWDHARY*

School of Information Technology and Engineering,
Vellore Institute of Technology, Vellore, India

**Corresponding author. E-mail: c.l.chowdhary@gmail.com*

ABSTRACT

Chatbot is a vivid human-like contrivance which bounces chat. The main impartial of the chatbot is to authorization the customers and the machines to associate each other to altercation their chats. The way a machine can know human chats and how they response to the clients is a stimulating effort. Chatbot by means of neural-network and deep learning models give tremendously good consequences in carrying a human verbal exchange. There are many present chats models; however, they have some problems. In this chapter, we discuss about sequence-to-sequence neural-network model which ends up some responses by training with datasets. So, a bot can hardly replace a human, but it is a great help to accomplish specific objectives with a limited reach. Receding from the general use chatbots are given, a useful product can be obtained, one that allows clients to have a different experience without feeling plagued with useless and senseless information.

12.1 INTRODUCTION

Chatbot is a brilliant human-like machine which gives conversation.[2] The primary objective of the chatbot is to permit the clients and the machines to connect each other to exchange their conversations. The way a machine can recognize human conversations and how they answer to the clients is a challenging work. Chatbot utilizing sequence-to-sequence show is a developing approach which provides a human-like conversation. This model builds a neural network that peruses a sentence and gives an extraordinary response[4] The sequence-to-sequence-based model grants a fruitful comes about in sequence mapping process. These shows have made an awesome progress in language translation, sentence generation system, and in speech to text.[12] It is first proposed to language translation, but also utilized in producing conservation as well. The sequence-to-sequence system has two main modules: encoder and decoder. An encoder neural network peruses the source sentence and encodes it to vector. A decoder at that point produces a response message from those vectors.[10,16–25] The conservation model has two models: generative-based and retrieval-based. The retrieval-based model has certain restrictions; they have predefined responses stored in database and they do not produce any new responses. But the generative-based show makes an unused reaction from scratch. The chatbot can be both closed and open domain. In open domain, there is no aim or purpose. It is like conservation on social network could be on any subject matter and there is no restriction on the barriers. In closed domain, the conservation is limited and the system need to acquire unique goals like personal assistance, customer care which are the best examples for closed domain chatbot.

Chatbots are also used in twitter social network for entertainment and for advertisement. These bots use twitter datasets and provide a realistic conversation like human. Using the chatbot for advertisement is really a good idea which increases the productivity of the product. The drawback is of identifying specific product details from twitter is really a complex task.[2] A hybrid neural network model which contains of some important network model. Here, the datasets are trained and experimented results show that the best accuracy belongs to different hybrid model. The chatbots using more than one neural network for can make our work easier and the output will be more accurate. If alignment between user input and output is certain then RNN can easily map the input sentence to the source sentence. Using more than one neural networks is really time-consuming and need more computations power.[3] Conservation model is important task in artificial intelligence as it understands the question and reply for that question. In this paper the conservation model is built by using sequence-to-sequence framework. Replies are generated by training large datasets. Conservation model can be open domain or closed domain. The conservation model can vary based on the datasets example IT helpdesk datasets can be used to provide solutions for technical problem. Predefining the dataset for certain question will give a better response. Rule-based chatbots are outdated and they provide a linear conservation. All the responses are stored in database and only response to specific questions[4] provides an outline of build neural-based conservation. Here, they used machine translation using neural network which help in increase the translation performance. These networks consist of encoder and decoder which help in translation input into source destination output. Building a neural-network model helps in increase the translation process.

Chatbot using neural-network model gives extremely good outcomes in delivering a human verbal exchange. There are numerous present conservation models; however, they have some obstacles. Here we discuss about sequence-to-sequence neural-network model which ends up some responses by training with datasets. Sequence-to-sequence model have two primary modules encoder and decoder. The encoder typifies the target sentence to a middle representation. Then decoder uses that representation and creates destination sentence. This model is applied for open and closed domain. The model can learn information from datasets. Utilizing domain specific dataset can be used as solutions for certain issues.

12.2 THE PROPOSED MODEL

12.2.1 RECURRENT NEURAL NETWORK

It is one among the deep learning concepts. This model is beneficial with sequential generation because it uses memory units to keep previous input data. This helps neural network to gain deeper information of the sentences. The RNN model acts a loop that permits the data to be carried throughout the layers inside them (Fig. 12.1).

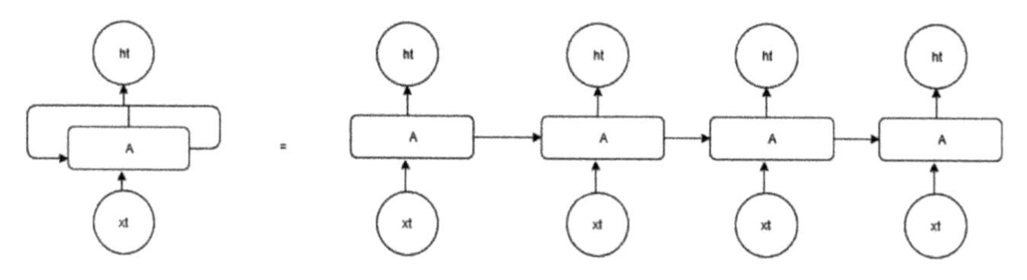

FIGURE 12.1 RNN.

In this above figure x_t is input, A is part of RNN and ht is the output. The main purpose is to get h_t as an output and examine it with the test data. From the output we can be getting an error rate. By comparing the error function with test dataset, we update the weights; this technique is called back propagation through time. By using the BPTT we can look through network and modify the weight to make the network learn better.

RNN model can handle many language model and extensively utilized in many sequence generation systems such speech and language translation this model also utilized in conversation generation by training the datasets and predict those conversations. In practice, the achievement of RNN is obtained by LSTM which work as a memory unit inside the neural network helps in keep remembering the facts from long time period.

12.2.2 LONG SHORT TERM MEMORY

LSTM is unique form of RNN which holds records for long time. LSTM have become very popular and are used in many problems. They are utilized in long term dependency problems (Fig. 12.2).

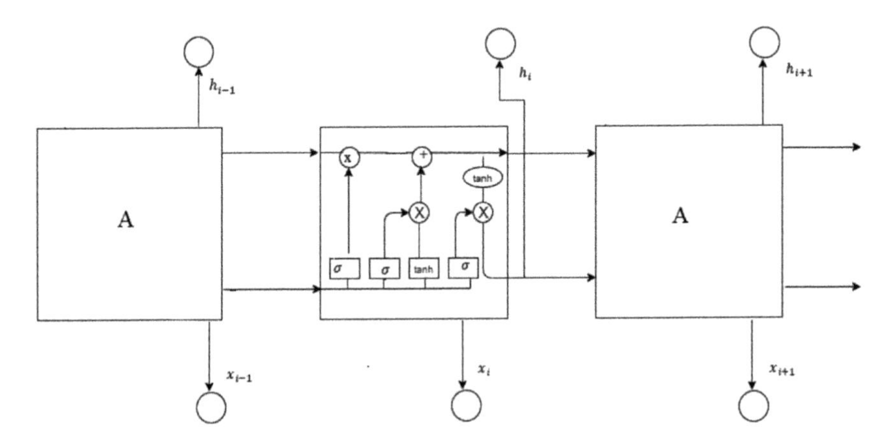

FIGURE 12.2 LSTM RNN cell.

The RNN model carries out repeating module, works as a loop and follows a simple structure. LSTM uses four interactive functions. LSTM cell consist of a conveyor belt the run across whole layers. It is easy to pass information along the layers. The LSTM have no power to add or to remove values in the cell states. They are managed by using operators known as gates. The most effective way to pass information is through gate. They designed out of sigmoid layer and point wise operation. It holds value range from zero to one. If value is 1 "allow everything through" if 0 it "allow nothing through."

Step one is to decide what information that we are going to take away. This is done by using sigmoid layer. It takes h_i−1 and in_t as input then the layer throws a number between 0 and 1 for every number within each cell state Ca_t−1. 1 represents "to preserve this information" while 0 represents "eliminate the information."

$$ft = \sigma(We_f\left[h_i - 1, in_t\right] + bf)$$

The following step is to decide which data is going to be kept within the cell state. It has two layers. A sigmoid layer as "input layer," it will decide which data to be replaced and a tanh layer which is responsible for introducing new candidate value $\check{Ca}_t$. After that integrate these values to generate new value.

$$it = \sigma(Wei.\left[h_{i-1}, in_t\right] + bi$$

$$\check{Ca}_t. = \tanh(We_c.[h_{i-1}, in_t] + bc$$

In the next layer, the old cell state $\check{Ca}_t$ Ca needs to be updated into new state Ca_t. Multiply old state with fo_t for updating the old cell state. Later, multiply $it * \check{Ca}_t$ to get new cell state.

$$Ca_t = fo_t * Ca_{i-1} + it * \check{Ca}_t$$

In the end, output of the LSTM is based on the cell state. Pass the value to sigmoid layer that decides the output, then the cell state is send through tanh layer. Later, multiply cell state by outcome of the sigmoid gate, then our output is generated.

$$ot = \sigma(We_o\left[h_i - 1, in_t\right] + b_o)$$

$$hi_t = ot * \tanh(Ca_t)$$

NN provides an excellent result these are achieved using LSTM. Here, we are using LSTM-based model for developing a conversation chatbot. The model has encoder and decoder helps in matching the input to output sentence.

12.3 DATASETS

The movie corpus dataset is used in this research work. This dataset consists of rich collection of realistic conservations which are derived from various movies. It consists of 245,579 conservations exchange between 10,457 pairs of movie character and involves 9345 characters from 787 movies. The dataset is stored in JSON format which involves character id, character name, movie id, movie title, line id, movie id, character id of 1st person, and character id of 2nd person. This corpus dataset has large pair of phrases and natural flow in the speech. This dataset is used for conservation generation. For the conservation bot we need dataset like questions and answers (i.e., sequences of questions and exact answers for those questions. Before passing data to the model the dataset need to organize.

12.4 PROCESSING

The RNN model does not take the input in strings; we have to convert the strings into ids. The movie conversation that we are passing into the network should be of equal size. So we have to pad for the both encoder and decoder inputs.

Finding out max sequence length is the primary task. For encoder input, the pad variable use for padding at the starting of the sequence with <BOS> tags and for decoder the padding variable <EOS> is used at end of the sentence. After padding is done we tokenize every sentence after convert each sentences into token of words. After, we convert each word in to unique ids.

12.5 TRAINING

The training process is chosen first to train the dataset. The processed encoder input and decoder input and batch size are passes as input for the RNN. Tensor flow class provide various RNN cells, here LSTM cell is used which is useful to hold information for long period of time. Next step provides the hidden size for the model, here hidden size is 512 in each RNN cell. This value in optional and it depends upon our model. Choosing lower value will provide poor result and too high values will over fit the model. The RNN cell generates a dictionary variable as output consists of decoder output and hidden states. After define weight and biases and multiply them with the output. After multiplying the output, the softmax function is calculated to generate the probability value. The probability value is the output value that we expect. Later find the loss function that is, incorrectness of training which is used as our cost function. It helps in finding out how well our data are trained against the actual results. This function is repeated to reduce our loss. The loss function helps in enhance the model. The final work is to develop an optimizing function.

Tensor flow supports many optimizing functions like RMSProp Optimizer, AdaGrad Optimizer, and so on. We run optimizer feature to reduce the cost. All the estimation and evaluation for predicting cost, and backpropagation is take care via tensor flow.

12.6 RESULTS

The network is much slower to train. After 19,000 iterations the networks try to provide a reply for some question but for most cases the model reply with noisy answers. The model can be improved by increasing the number of iteration. In Figure 12.3, the loss functions get gradually decreased to improve the learning rate.

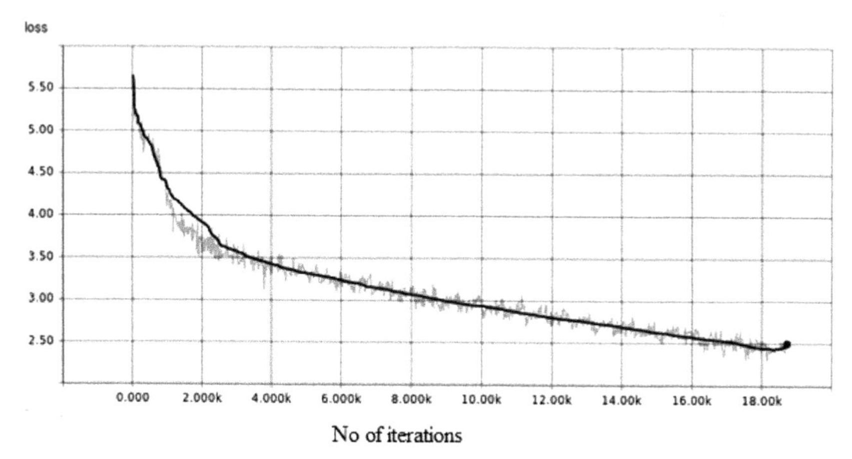

FIGURE 12.3 (See color insert.) Loss minimizer.

TABLE 12.1 Questionnaire and Answers.

S. no.	Questions	Answers
1	Hi	Hi
2	Hello	Hi
3	How are you?	Fine
4	Where are you from?	Southern California
5	How old are you?	Twenty-five
6	When is you birthday	Tomorrow
7	I am john	Oh. Hi.
8	What is your name?	Bedevere, my liege
9	Have we met before?	No
10	Do you have any friends?	Yes
11.	Well, shall we dance	I am also nervous
12	I am ill	What's wrong?
13	What's wrong?	Oh?
14	Can you cure me	I'm sorry
15	Please help me	: But ...
16	But?	: No buts
17	Have you met god?	I tried to go back to the southland
18	Where do you see him?	At the Hilton
19	Bye	Yes, yes!
20	See you	Bye

12.7 CONCLUSIONS AND FUTURE WORKS

A bot can hardly replace a human, but it is a great help to accomplish specific objectives with a limited reach. Receding from the general use chatbots are given, a useful product can be obtained, one that allows clients to have a different experience without feeling plagued with useless and senseless information. The next is to improve the performance of these bots, besides phrase or word hierarchy, is adding a numeric method to understand content of the message, to distinguish the mood and the sense of conversation, leaving aside grammatical errors. Such grammatical errors generally are not considered and make the final user think the bot is programmed incorrectly, or that it just does not work the way it should. It would be of great help being able to add other data to the tables of the database, to have more information that allows us to select an answer more efficiently. Leaning on the option Twitter offers to look a conversation history, an extract field could be defined, containing the IDs of the answers that should have previously been sent, so the selected answer can turn as valid. This way, a context of what the heading of conversation is, and what path conversation is going, can be obtained. To design chatbot, various parameters need to consider. It is essential to monitor continuously its operation at the initial level and, if necessary, make the appropriate changes. This makes possible decrease time to approach with the bot and do not have opportunity to know all its answers or limitations.

KEYWORDS

- **chatbot**
- **deep learning**
- **RNN**
- **LSTM**
- **BPTT**
- **AdaGrad optimizer**

REFERENCES

1. Bahdanau, D.; Cho, K.; Bengio, Y. *Neural Machine Translation by Jointly Learning to Align and Translate*; ICLR, 2015; pp 1–15.
2. Sutskever, I.; Vinyals, O.; Le, Q. V. *Sequence to Sequence Learning with Neural Networks*; NIPS: Montreal, CA, 2014; pp 1–9.
3. Rodrigo, S. M.; Abraham, J. G. F. In *Development and Implementation of a Chat Bot in a Social Network.* Proceedings 9th International Conference on Information Technology, ITNG, 2012; pp 751–755.
4. Vinyals, O.; Le, Q. *A Neural Conversational Model*, ICML Deep Learning Workshop, 2015; p 37.
5. Serban, I. V.; Sordoni, A.; Bengio, Y.; Courville, A.; Pineau, J. *Building End-To-End Dialogue Systems Using Generative Hierarchical Neural Network Models*, AAAI 2016 (Special Track on Cognitive Systems), 2016.

6. Li, J.; Galley, M.; Brockett, C.; Gao, J.; Dolan, B. In *A Diversity-Promoting Objective Function for Neural Conversation Models*. Proceedings of NAACL-HLT 2016, San Diego, California, 2016; pp 110–119.
7. Yao, K.; Zweig, G.; Peng, B. *Attention with Intention for a Neural Network Conversation Model*. ArXiv, 2015; pp 1–7.
8. Li, J.; Galley, M.; Brockett, C.; Spithourakis, G. P.; Gao, J.; Dolan, B. In *A Persona-Based Neural Conversation Model*. Proceedings of the 54th Annual Meeting of the Association for Computational Linguistics, Berlin, Germany, 2016; pp 994–1003.
9. Gu, J.; Lu, Z.; Li, H.; Li, V. O. K. In *Incorporating Copying Mechanism in Sequence-to-Sequence Learning*. Proceedings of the 54th Annual Meeting of the Association for Computational Linguistics, Berlin, Germany, 2016; pp 1631–1640.
10. Haller, E.; Rebedea, T. In *Designing a Chat-bot That Simulates an Historical Figure*. Proceedings of the 19th International Conference on Control System and Computing Science CSCS, 2013; pp 582–589.
11. Mathur, V.; Stavrakas, Y.; Singh, S. In *Intelligence Analysis of Tay Twitter Bot*. Proceedings of the 2016 2nd International Conference on Contempt Computing Informatics, IC3I 2016, pp 231–236.
12. Sun, X.; Peng, X.; Ren, F.; Xue, Y. In *Human-Machine Conversation Based on Hybrid Neural Network*. IEEE International Conference on Computing Science and Engineering. IEEE International Conference Embedded Ubiquitous Computing. **2017,** *1*, 260–266.
13. Fujita, T.; Bay, W.; Quan, C. *Long Short-Term Memory Networks for Automatic Generation of Conversations*, 2017; 483–487.
14. Ceresin, D.; Meier, U.; Schmidhuber, J. In *Multi-column Deep Neural Networks for Image Classification*. CVPR, 2012.
15. Papineni, K.; Roukos, S.; Ward, T.; Zhu, W. J. In *BLEU: A Method for Automatic Evaluation of Machine Translation*. ACL, 2002.
16. Chowdhary, C. L.; Acharjya, D. P. In *Singular Value Decomposition–Principal Component Analysis-Based Object Recognition Approach*. Bio-Inspired Computing for Image and Video Processing, 2018; p 323.
17. Chowdhary, C. L. In *Application of Object Recognition With Shape-Index Identification and 2D Scale Invariant Feature Transform for Key-Point Detection*. Feature Dimension Reduction for Content-Based Image Identification, 2018; pp 218–231.
18. Chowdhary, C. L.; Muatjitjeja, K.; Jat, D. S. In *Three-Dimensional Object Recognition Based Intelligence System for Identification*. Emerging Trends in Networks and Computer Communications (ETNCC), 2015.
19. Chowdhary, C. L.; Ranjan, A.; Jat, D. S. Categorical Database Information-Theoretic Approach Of Outlier Detection Model. *Ann. Comput. Sci. Ser.* **2016,** *2016*, 29–36.
20. Chowdhary, C. L. Linear Feature Extraction Techniques for Object Recognition: Study of PCA and ICA. *J. Serb. Soc. Comput. Mech.* **2011,** *5* (1), 19–26.
21. Chowdhary, C. L.; Acharjya, D. P. In *Breast Cancer Detection Using Hybrid Computational Intelligence Techniques*. Handbook of Research on Emerging Perspectives on Healthcare Information Systems and Informatics, 2018; pp 251–280.
22. Chowdhary, C. L.; Acharjya, D. P. In *Segmentation of Mammograms Using a Novel Intuitionistic Possibilistic Fuzzy C-Mean Clustering Algorithm*. Nature Inspired Computing, 2018; pp 75–82.

23. Chowdhary, C. L.; Acharjya, D. P. Clustering Algorithm in Possibilistic Exponential Fuzzy C-Mean Segmenting Medical Images. *J. Biomimet. Biomater. Biomed. Eng.* **2017,** *30,* 12–23.
24. Chowdhary, C. L.; Acharjya, D. P. In *A Hybrid Scheme for Breast Cancer Detection Using Intuitionistic Fuzzy Rough Set Technique*. Biometrics: Concepts, Methodologies, Tools, and Applications, 2016; pp 1195–1219.
25. Das, T. K.; Chowdhary, C. L. Implementation of Morphological Image Processing Algorithm Using Mammograms. *J. Chem. Pharm. Sci.* **2016,** *10* (1), 439–441.
26. Chowdhary, C. L. A Review of Feature Extraction Application Areas in Medical Imaging. *Int J. Pharm. Technol.* **2016,** *8* (3), 4501–4509.
27. Chowdhary, C. L.; Acharjya, D. P. In *Breast Cancer Detection Using Intuitionistic Fuzzy Histogram Hyperbolization and Possibilitic Fuzzy c-mean Clustering Algorithms With Texture Feature Based Classification on Mammography Images.* Proceedings of the International Conference on Advances in Information Communication Technology & Computing, 2016; p 21.
28. Chowdhary, C. L.; Sai, G. V. K.; Acharjya, D. P. Decreasing False Assumption for Improved Breast Cancer Detection. *J. Sci. Arts* **2016,** *35* (2), 157–176.
29. Chowdhary, C. L.; Sai, G. V. K.; Acharjya, D. P. Decrease in False Assumption for Detection Using Digital Mammography. *Comput. Intell. Data Mining* **2015,** *2,* 325–333.
30. Chowdhary, C. L. *Appearance-based 3-D Object Recognition and Pose Estimation: Using PCA, ICA and SVD-PCA Techniques*; LAP Lambert Acad, 2011.

PART III

Intelligent Systems and Hybrid Systems

CHAPTER 13

A Real-Time Data Analytics-Based Crop Diseases Recognition System

CHIRANJI LAL CHOWDHARY*, RACHIT BHALLA, ESHA KUMAR, GURPREET SINGH, K. BHAGYASHREE, and GURSIMRAN SINGH

Department of Software and System Engineering, School of Information Technology and Engineering, Vellore Institute of Technology, Vellore 632014, India

**Corresponding author. E-mail: c.l.chowdhary@gmail.com*

ABSTRACT

This chapter is focused on the use of deep learning concepts and analyzing real-time data based on the location of users to detect effectively and rapidly crop diseases and pests. The images and other data like locations will be collected using camera devices like mobile and sent to an online server. The server will filter the possible diseases and pests by using previously collected data and downloaded data related to a given location like temperature conditions. Then, neural networks can be used to figure out crop diseases/pests and results would be sent back to devices.

13.1 INTRODUCTION

The whole world is reliant on agriculture where farmers are the customary victims of crop diseases and pests, occasionally even deprived of knowing what the disease is. If researchers can provide some approaches to recognize and detect the diseases at an aforementioned juncture and arise with an elucidation for it, the total productivity can be improved and monetarist losses can be reduced. The agriculture should be harmless, healthy, and maintainable; it is indispensable to have healthy crops. They play a role in producing enough amounts of healthy foods and subsidize to the excellence of lifespan. The research work into disease and pest supervision for the agricultural and horticultural segments emphases on the collaboration among crops and the diseases and pests that affect them.

Assimilated crop fortification comprises uniting numerous sustainable crop protection approaches in edict to elude diseases and pests or to overturn them. The intention is to reduce harm to the environment. Chemical agents are used only to a precise range.

The usage of unified crop protection approaches so as to mark the agricultural and horticultural segments fewer reliant on pesticides. Cohesive crop control should contain some stages. Formation of agronomy approaches and selected plants should encompass precautionary measures such as by means of disease-free seed, choosing robust variations, and organizing resilient schemes. Throughout the growing retro, the crop should be watched prudently by means of, for example, simulation models and decision support systems. If a disease or pest does loom to touch a crop, the control technique should be designated by upkeep. If the lone choice is a pesticide, a biological one is favored—composed by powered and additional nonchemical systems. Chemical resources had better to be castoff merely as a former option.

As deep learning has made it conceivable to perceive objects exhausting images abundant additional efficiently, the researcher may improve the matching skill to undertake crop diseases recognition objective. In this chapter, authors focus on the use of deep learning concepts and analyzing real-time data based on the location of users to detect effectively and rapidly crop diseases and pests. The images and other data, like location, will be collected using camera devices like mobile and sent to the online server. The server will filter the possible diseases and pests by using previously collected data and downloaded data related to a given location like temperature conditions. Then, neural networks can be used to figure out crop diseases/pests and result would be sent back to devices.

13.2 OBJECTIVES OF CROP DISEASES RECOGNITION SYSTEM

- The main objective of the crop image is to detect the disease affected areas and recognize the pest/disease.

- The database associated with effective cure of diseases can also be stored. So that farmers can be provided with a reliable solution to diseases.
- Our system uses images of plant diseases and pests taken in-place, thus we avoid the process of collecting samples and analyzing them in the laboratory.
- It considers the possibility that a plant can be simultaneously affected by more than one disease or pest in the same sample. Our approach uses input images captured by different camera devices with various resolutions, such as cell phone and other digital cameras.
- Even the government can use this data to figure out specific disease/pest-affected areas and provide required chemicals (pesticides) at subsidy to the farmers. Moreover, new diseases/pests for which neural-networks failed to identify distinctively can be tracked and hence scientists may start finding their cure at the earliest.

Plant diseases visibly show a variety of shapes, forms, colors, etc. Understanding this interaction is essential to design more robust control strategies to reduce crop damage. Moreover, the challenging part of our approach is not only in disease identification but also in estimating how precise it is and the infection status that it presents.

Several methods have been proposed earlier to detect crop disease/pest recognition. Most of these works like "Deep Neural Networks Based Recognition of Plant Diseases by Leaf Image Classification (Computational Intelligence and Neuroscience Volume 2016)" focused on the use of convolutional neural networks and use of augmentation to increase the dataset which gave an accuracy of 94.67%.

The most effective method is proposed in "A Robust Deep-Learning-Based Detector for Real-Time Tomato Plant Diseases and Pests Recognition (Sensors Nov. 2017)" the combined use of detectors (Faster Region-based Convolutional Neural Network, Region-based Fully Convolutional Network, and Single Shot Multibox Detector) and feature extractors that gave much more accuracy.

We can improve this framework combined with extractors by including real-time data analysis (location, region-based disease, crop-disease pairs) for accurate and quick diseases and pests recognition.

The commitments of this chapter are as per the following:

- A vigorous profound learning-based detector is proposed for constant illnesses and pest identification.
- The framework presents a functional and pertinent answer for recognizing the class and area of diseases, which in certainty speaks to a fundamental practically identical distinction with conventional techniques for plant maladies characterization.
- This detector makes use of pictures caught set up by different camera gadgets that are prepared by real-time hardware and software system, programming framework utilizing graphical preparing units (GPUs), instead of utilizing the procedure of gathering physical examples (leaves, plants) and dissecting them in the lab.
- Moreover, it can productively manage distinctive work complexities, for example, light conditions, the size of objects, and foundation varieties contained in the encompassing territory of the plant.

13.3 LITERATURE REVIEW

Several techniques have been recently applied to identify plant diseases. These include using direct methods closely related to the chemical analysis of the infected area of the plant, and indirect methods employing physical techniques, such as imaging and spectroscopy, to determine plant properties and stress-based disease detection. However, the advantages of our approach compared with most of the traditionally used techniques are based on the following facts:

- The system uses images of plant diseases and pests taken in-place, thus we avoid the process of collecting samples and analyzing them in the laboratory.
- It considers the possibility that more than one disease or pest in the same sample can simultaneously affect a plant.
- This approach uses input images captured by different camera devices with various resolutions, such as cell phone and other digital cameras.
- It can efficiently deal with different illumination conditions, the size of objects, and background variations, etc., contained in the surrounding area of the plant.
- It provides a practical real-time application. Recent advances in hardware technology have allowed the evolution of deep convolutional

neural networks and their large number of applications, including complex tasks, such as object recognition and image classification.

Here, diseases are addressed and pest identification by introducing the application of deep meta-architectures and feature extractors. Instead of using traditionally employed methods, we develop a system that successfully recognizes different diseases and pests in images collected in real scenarios. Furthermore, our system is able to deal with complex tasks, such as infection status (e.g., earlier, last), location in the plant (e.g., leaves, steam), sides of leaves (e.g., front, back), and different background conditions, among others.

The following characteristics are usually considered for analysis:

- Infection status: A plant shows different patterns along with their infection status according to the life cycle of the diseases.
- Location of the symptom: It considers that diseases not only affect leaves, but also other parts of the plant, such as stem or fruits.
- Patterns of the leaf: Symptoms of the diseases show visible variations either on the front side or the backside of the leaves.
- Type of fungus: Identifying the type of fungus can be an easy way to visibly differentiate between some diseases.
- Color and shape: Depending on the disease, the plant may show different colors or shapes at different infection stages.

A few strategies have been as of late connected to clearly distinguish plant sicknesses. These incorporate utilizing direct strategies firmly identified with the chemical analysis of the infected area of the plant, and indirect strategies utilizing physical systems, for example, imaging and spectroscopy, to decide plant properties and stress-based infection location. In any case, the benefits of our approach contrasted with the traditional procedures depend mainly on the following reasons:

- The framework utilizes pictures of plant maladies and pests assumed in-position; in this manner, we maintain a strategic distance from the procedure of gathering tests and dissecting them in the research center.
- It thinks about how conceivable it is that a plant can be at the same time influenced by more than one illness or pest in a similar example.
- This approach utilizes input pictures caught by various camera gadgets with different resolutions, such as cell phone and other advanced cameras.

- It can productively manage diverse light conditions, the measure of items, and foundation varieties, and so on, contained in the encompassing territory of the plant.
- It gives a handy practical real-time application.
- Late advances in equipment innovation have evolved the development of Deep Convolutional.

Neural Networks and their vast number of uses, including complex undertakings, for example, as object recognition and image classification.

In this chapter, we address disease and pest identification by presenting the utilization of profound meta-structures and feature extractors. Rather than utilizing generally utilized techniques, we fundamentally build up a framework that effectively perceives diverse diseases and bugs in pictures gathered in genuine situations. Besides, the framework can manage complex errands, for example, contamination status, an area in the plant, sides of leaves, and diverse foundation conditions, among others.

13.4 INVENTION DETAILS

This is a list of characteristics for the analysis:

- Infection status: A plant shows different patterns along with their infection status according to the life cycle of the diseases.
- Location: Inclusion of location of the user for faster identification of disease based on data analysis.
- Type of fungus: Identifying the type of fungus can be an easy way to visibly differentiate between some diseases.
- Color and shape: Depending on the disease, the plant may show different colors or shapes at different infection stages.

Plants are susceptible to several disorders and attacks caused by diseases and pests. If we are able to detect diseases accurately and efficiently, we may save time and money. So, the proposal includes the inclusion of location of crop field for faster analysis of diseases based on data analysis, weather conditions, etc.

13.5 ARCHITECTURE OF PROPOSED WORK

13.5.1 FEATURE EXTRACTORS

Hu-moment, Harlick-texture, color histogram, Resnet model

13.5.2 ACTIVATION FUNCTION

Activation function used is "ReLU," it is the most effective function for efficient result till date. It is stated as "Rectified Linear Unit" and approximation to $f(x)=\log(1+\exp(x))$ which is soft plus function. Its derivative is given by $f'(x)= \exp(x)/(1+\exp(x))$.

13.5.3 INTERCONNECTION

Interconnection between a web app and python is done using flask library. It connects the web app function with python functions and returns the result on the web page.

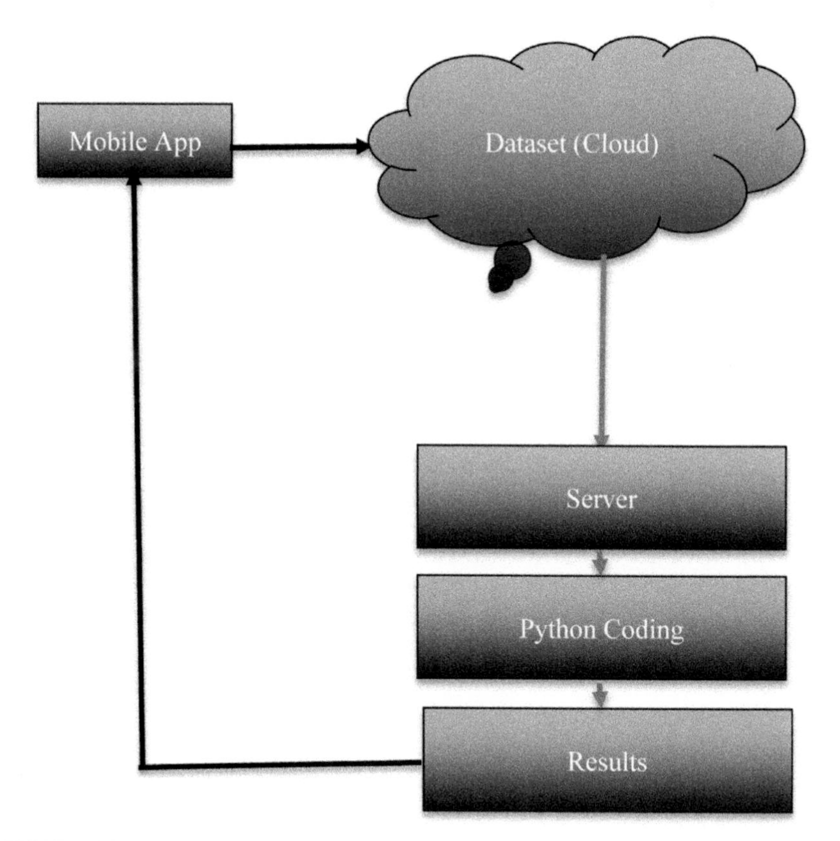

FIGURE 13.1 Dataset training overview: testing flow of images.

13.5.4 BACK END

Being one of most of the useful and advanced language in terms of machine learning, back-end parts have been done using python. All the training and testing is being done in python.

13.5.5 FRONT END

Front-end part is made using Android Studio for making a mobile app and html/css/angular for making the web app.

13.6 DETAILED DESCRIPTION OF THE INVENTION

Our work aims to identify types of diseases and pests that affect plants using Deep Learning as the main body of the system. Following we describe in detail (Table 13.1) each component of the proposed approach (Fig. 13.2).

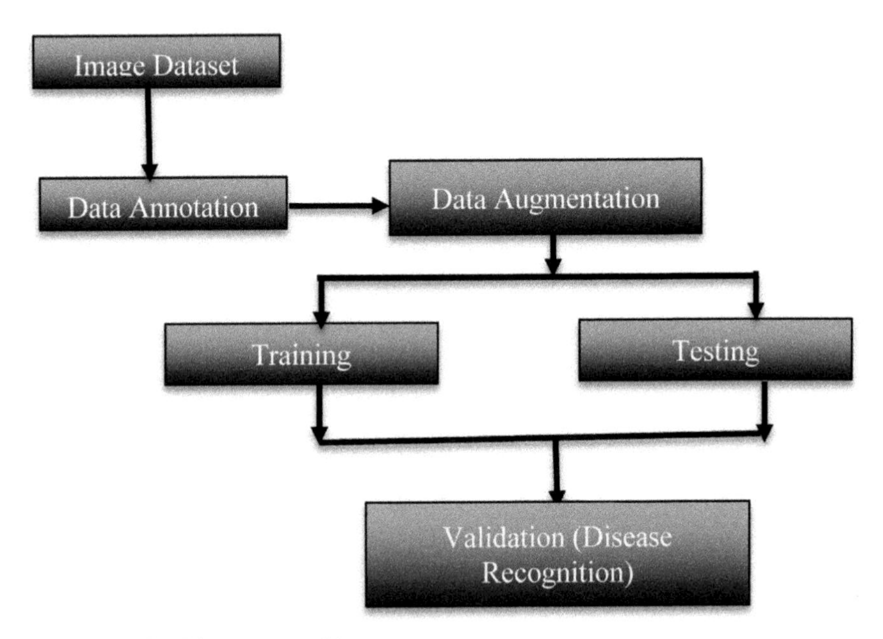

FIGURE 13.2 Disease recognition steps.

13.6.1 DATA COLLECTION

Our dataset contains the images collected from various sources.

13.6.2 DATA ANNOTATION

Since our images are collected in the field, many areas corresponding to the background could be included in the image, making the problem more challenging. Therefore, when collecting the images, we find out that the best way to get more precise information is to capture the samples containing the "leaf" as the main part of the image.

13.6.3 DATA AUGMENTATION

Due to lack of dataset and increasing efficiency, it is important to have data augmentation. Data augmentation is necessary to pursue when the number of images in the dataset is not enough. We use several techniques that increase the number of images of our dataset. It includes geometrical transformations (resizing, cropping, rotating, and horizontal flipping).

13.6.4 DISEASE AND PEST DETECTION

We now describe our main method for detecting diseases and pests. Our goal is to detect and recognize the class and pest candidates in the image. To detect our target, we need to accurately localize the box containing the object as well as identify the class to which it belongs.

13.7 WORKING SCENARIO

First of all, the image is sent to online sever (firebase) after capturing it using a mobile phone camera using the app.

Then the python code is fetched from online storage server (firebase) and processed to detect the disease.

After the disease is detected, it is sent to the web page, where the user can check the details of pest/disease detected and its optimum treatment.

13.7.1 IMPLEMENTATION—SCREENSHOTS

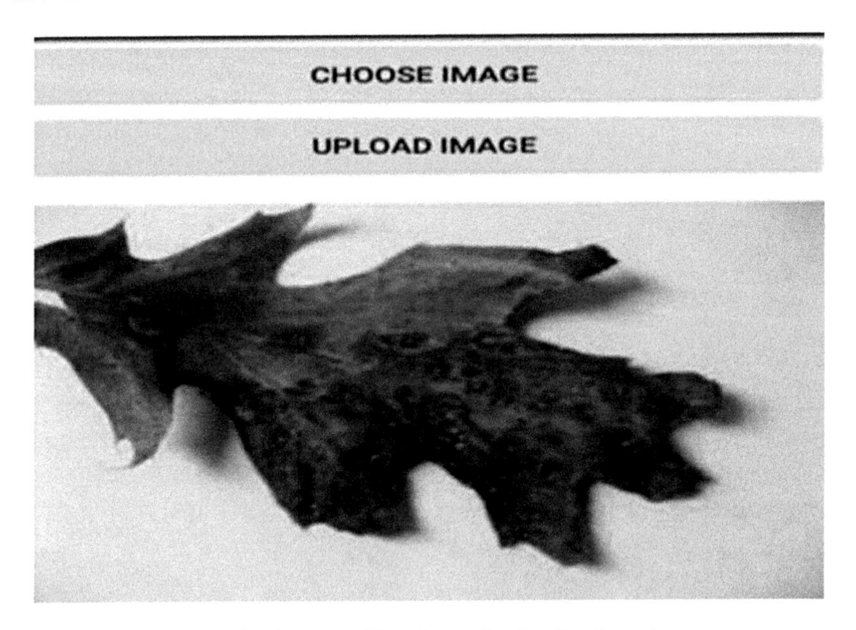

FIGURE 13.3 **(See color insert.)** Choosing and uploading input image.

Farming House

Diseases:

- [illegible]

Treatment:

[illegible]

[illegible]

FIGURE 13.4 **(See color insert.)** Identifying disease and sharing details.

13.8 CONCLUSIONS

In this work, a new approach of using deep learning method was explored in order to automatically classify and detect plant diseases from images. The developed model was able to detect leaf presence and distinguish between healthy leaves and diseased leaves, which can be visually diagnosed. The complete procedure was described, respectively, from collecting the images used for training and validation to image preprocessing and augmentation and finally the procedure of training. Different tests were performed in order to check the performance of the newly created model. New plant disease image database was created.

Here, we have proposed a robust deep-learning-based detector for real-time diseases and pests recognition. This system introduces a practical and applicable solution for detecting the class and location of diseases in plants. Our detector applied images captured in-place by various camera devices and processed them by a real-time hardware and software system using GPUs, rather than using the process of collecting physical samples (leaves, plants) and analyzing them in the laboratory. Furthermore, the plant diseases and pest dataset contains different task complexities, such as illumination conditions, the size of objects, background variations, etc., included in the surrounding area of the plant. Our goal was to find the more suitable deep-learning architecture for our task. Thus, the experimental results demonstrated how our deep-learning-based detector is able to successfully recognize diseases and pests, including complex intra- and interclass variations. In addition, we found that using technique-based data annotation and augmentation results in better performance. We expect that our proposed system will make a significant contribution to the agriculture research area.

KEYWORDS

- **Deep Neural Network**
- **GPU**
- **deep learning**
- **real-time data analytics**
- **crop diseases recognition system**

REFERENCES

1. www.ncbi.nlm.nih.gov/pmc/articles/PMC5620500/
2. www.hindawi.com/journals/cin/2016/3289801/
3. www.frontiersin.org/articles/10.3389/fpls.2016.01419/full
4. https://arxiv.org/pdf/1604.03169
5. www.researchgate.net/publication/301879540_Using_Deep_Learning_for_Image-Based_Plant_Disease_Detection
6. www.europepmc.org/abstract/med/27713752
7. europepmc.org/abstract/med/27713752
8. https://en.wikipedia.org/wiki/Lists_of_plant_diseases
9. https://www.daf.qld.gov.au/plants/field-crops-and-pastures/broadacre-field-crops/integratedpest-management/a-z-insect-pest-list
10. https://images.google.com/
11. https://www.crowdai.org/challenges/plantvillage-disease-classification-challenge
12. ttps://www.apsnet.org/publications/imageresources/Pages/default.aspx
13. http://leafsnap.com/dataset/

CHAPTER 14

Image Caption Generation with Beam Search

CHIRANJI LAL CHOWDHARY*, AMAN GOYAL, and
BHAVESH KUMAR VASNANI

*Department of Software and System Engineering,
School of Information Technology and Engineering,
Vellore Institute of Technology, Vellore 632014, India*

Corresponding author. E-mail: c.l.chowdhary@gmail.com

ABSTRACT

The image caption generation is a major problem faced in many situations. It basically means to generate a descriptive sentence of the image. This makes it very complex as the machine has to deep learn from the datasets and then describe the objects, activities, and the places. The fact that humans can do it quite easily for small sets but fail when the number of images is more, make it a quite interesting challenge of deep learning. Many of the implementations are quite complex and some even require multiple passes. The image caption generation task can be simplified with the help of deep learning concepts of deep neural networks. The show and tell algorithm that incorporates the concept of LSTMs proves to be beneficial in simplifying the task to generate captions.

14.1 WHAT IS IMAGE CAPTION GENERATION?

Image caption generation is a major problem faced in many situations. It basically means to generate a descriptive sentence of the image. This makes it very complex as the machine has to deep learn from the datasets and then describe the objects, activities, and the places. The fact that humans can do it quite easily for small sets but fail when the number of images is more, make it a quite interesting challenge of deep learning. Many of the implementations are quite complex and some even require multiple passes.

Our approach for the image caption generation would be basically based on understanding long-short-term memory networks (LSTM) and recurrent neural networks (RNN). This network model allows us to select the next word of the sequence in a better manner. In image caption generation one platform offered is Python and with the help of TensorFlow library, user can easily generate LSTM model for the given images. For this purpose, first, train the machine by a dataset for deep learning. To improve the efficiency of the caption generation, the training has to be quite deep with more sample images.

In this chapter, a deep learning based on RNN for image caption generation with greater accuracy is discussed. This chapter would be able to generate quite appropriate captions for the images that can be given as inputs. The model would concatenate all the fragments of a word to give an output. The applications of the proposed approach are quite wide, for example, caption generation for vloggers who post quite a lot of photos daily, for news channels to get real-time captions with a single pass. Later, there is a comparison and analyses for the performance of this chapter.

14.2 OBJECTIVE OF IMAGE CAPTION GENERATION

14.2.1 CAPTION FOR DIFFERENT OBJECTS

Using this chapter, there is a classification of photos into mountains, sea, and so on which will help to cluster them and categorize them for better management of photos in any file management application.

14.2.2 REAL-TIME NEWS CAPTION

The newspaper company or news channel can use this project for faster generation of caption of news headlines for the photos which can help them to get ideas for better caption regarding the photos.

14.2.3 CAPTION FOR LIFELOGGING OR VLOGGING

Lifelogging cameras capture and save everyday life from a first-person perspective but at the same time they also produce enormous amount of data in the form of images and which are used in our automatic image captioning algorithms to generate textual representations of these collections to get a better idea about the image and directly post them with the generated caption immediately.

14.2.4 AIDING BLIND PEOPLE

Blind people can use this application with a little further advancement in it which can help them in recognizing the object in front of them by giving the audio output of the caption form the image and they can get an image in their mind.

14.2.5 MEDICAL IMAGE CAPTION FOR DIAGNOSIS

In medical sector, one can use this software to recognize the skin diseases so that they can get proper medication. Not only doctors but patients can also use it for getting the idea on the disease they might have.

14.3 LITERATURE SURVEY OF IMAGE CAPTION GENERATION

In the modern-day, with developments being made in the field of artificial intelligence, we see an increase in the number of problems from real-world scenarios that can be tackled with the help of deep learning. Automatic caption generation of images is one such application that is slowly being popularized in today's world. The main challenge behind this task is to efficiently combine artificial intelligence techniques involving both computer vision as well as parsing of natural languages. One such method that we wish to implement to efficiently combine these two fields is the implementation of a deep recurrent architecture that has been successfully utilized in existing image caption generators.[1]

A deep recurrent neural network is a type of artificial neural network that involves the units of the network being connected to each other in the form of a directed cycle, thus exhibiting temporal behavior of a dynamic nature wherein internal memory can be efficiently utilized to process any random input sequence, thus making this kind of network suitable for real-world applications like voice recognition and image captioning.[2]

To delve into the appropriate structuring of such a recurring network, we have researched on the two-directional mapping that is bound to be involved between the images themselves and the sentence based descriptions we wish to generate in with the help of this network. We aim to utilize a visual representation that automatically teaches itself to maintain memory of visual concepts in a long term perspective. An efficient way to modulize this huge process into many smaller tasks would be to segment it into sentence generation, sentence retrieval, and image retrieval. Existing implementations of such mapping such as seen in the Mind's Eye caption generator have resulted in the automatically generated annotations being preferred equally or over human-generated captions 21% of the time, which is quite a promising figure.[3]

One of the most relevant architectures currently being used for automatic image captioning is the CNN-LSTM architectures that has been frequently used on popular datasets such as Microsoft common objects in context (MCSCOCO) to generate captions across visual spaces. It was found that in most iterations, partial discrepancies were observed due to insufficient attention to specific image details and thus the chapter recommends incorporation of attention-mechanisms along with the LSTM architecture.[4]

Another field of survey to tackle the task of generation of annotations for images would be to classify the currently existing methodologies based on their point of views used in conceptualizing the problem, such as models that focus on image description as a retrieval problem versus the models that view image description as a generation problem across the visual representations or multimodal space. From our survey we have found that efficiently classifying the processes of automated image caption generation into: (1) generation of descriptions directly from the images, (2) retrieving images from a visual space, and (3) image retrieval from a multimodal space and combining these subprocesses with the help of our recurring neural network is the best existing method of generating annotations.[5,8–10]

For individuals to utilize various pictures adequately on the web, advances must have the capacity to clarify picture substance and must be fit for scanning for information that clients require. Besides, pictures must be depicted with normal sentences construct not just with respect to the names of articles contained in a picture yet additionally on their common relations. Su Mei Xi proposed a framework which produces sentential comments for general pictures. Right off the bat, a weighted component grouping calculation is utilized on the semantic idea bunches of the picture locales. For a given group, we decide pertinent highlights in view of their factual appropriation and appoint more noteworthy weights to significant highlights as contrasted with less significant highlights. Thusly the processing of bunching calculation can maintain a strategic distance from overwhelmed by trifling applicable or immaterial highlights. At that point, the connection between grouping districts and semantic ideas is set up as indicated by the marked pictures in the preparation set. Under the state of the new unlabeled picture districts, he ascertains the restrictive likelihood of each semantic Catchphrase and comment on the new pictures with maximal contingent likelihood. Investigations on the Corel picture set demonstrate the viability of the new algorithm.[6]

Another cascade recurrent neural network (CRNN) for picture subtitle generation is proposed. Unique in relation to the traditional multimodal recurrent neural network, which just uses a solitary system for removing unidirectional syntactic highlights, CRNN embraces a cascade network for taking in visual-dialect associations from forward and in reverse headings, which can abuse the profound semantic settings contained in the picture. In the proposed system, two implanting layers for thick word articulation are developed. Another stacked Gated Recurrent Unit is intended for learning picture word mappings. The viability of the CRNN display is confirmed with adopting the generally utilized MSCOCO data-sets, where the outcomes show CRNN can accomplish better execution contrasted and the best in class picture inscribing techniques, for example, Google NIC, multimodal repetitive neural system and many others.[7]

14.4 INVENTION DETAILS

14.4.1 OBJECTS OF THE INVENTION

The major objective is to generate real-time caption for any input image in a single pass and ensuring the accuracy of the result by training the machine with good datasets.

14.4.2 SUMMARY OF INVENTION

Consequently, portraying the substance of a picture is a key issue profound discovering that interfaces PC vision and regular dialect preparing. In this undertaking, we would introduce a generative model in view of a profound intermittent engineering that joins late advances in PC vision and machine interpretation and that can be utilized to produce characteristic sentences depicting a picture. We would make the machine to generate a suitable caption from the images based on the learning that we feed into it. Deep learning concept with the use of RNN would be quite useful and a big dataset would be required to provide learning to the machine. The process contains a series of sub-processes that would be learning through the data-sets, providing the input and then generating the caption from it. Ultimately, we are going to compare the accuracy in the results that are predicted from the two approaches one with beam search and another without beam search.

14.4.3 ARCHITECTURE OF THE CAPTION GENERATOR

In Figure 14.1, the words of the caption tried to predict followed by the word embedding vectors for each word. The outputs of the LSTM are probability distributions generated by the model for the next word in the sentence. The model is trained to minimize the negative sum of the log probabilities of each word. In the second approach, we added a beam search algorithm to increase the accuracy and compare the results (Fig. 14.2).

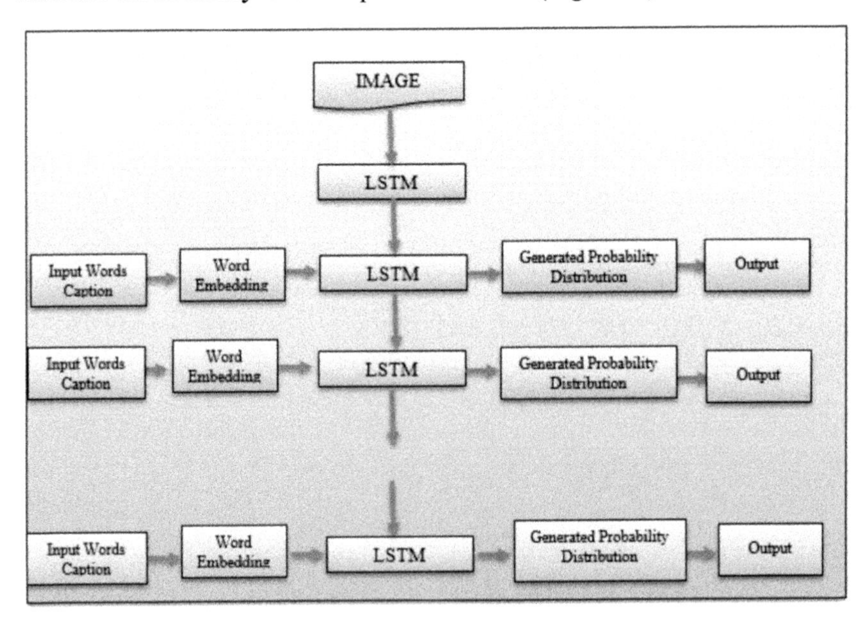

FIGURE 14.1 Architecture of caption generator.

Beam search is a heuristic finding algorithm that investigates a chart by growing the most encouraging hub in a restricted set. Beam search is an advancement of best-first hunt that diminishes its memory requirements. A beam search is regularly used to keep up tractability in huge frameworks with lacking measure of memory to store the whole inquiry tree. For case, it is utilized as a part of numerous machine interpretation frameworks.

14.4.4 DETAILED DESCRIPTION OF THE INVENTION

In this project, we are first taking the dataset from Flickr8k datasets. These datasets are then first passed through a feature capturing model which captures the feature of each image and start generating a model for producing a caption for the image. To generate the model, the program reads each image and its feature in details and tries to compare with other images if possible. These images have predefined caption with are given in the datasets and the model tries to read and learn from these captions and features of the images.

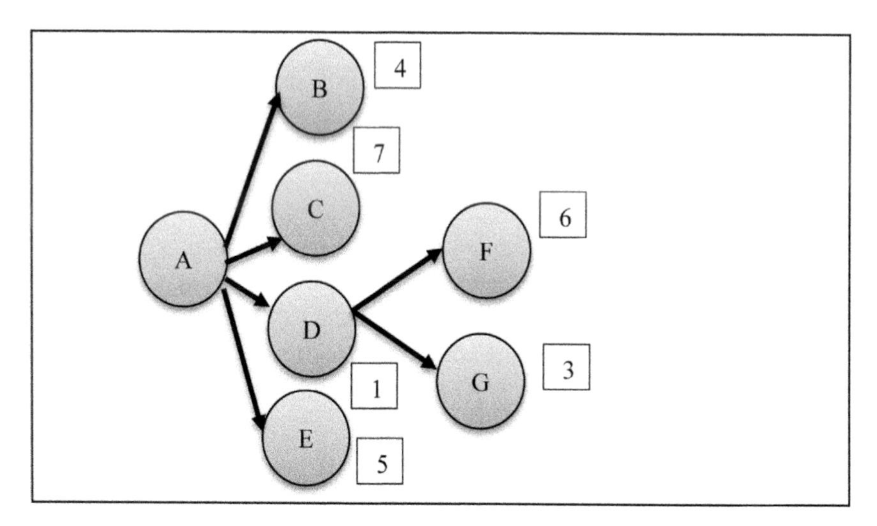

FIGURE 14.2 Beam search algorithm example step.

After training the dataset it sees the accuracy of the model by applying it to any random trained images. And then the model is ready for testing. This model is the given different varieties of images for testing whether the model is giving the proper output/caption for images or not.

14.4.5 WORKING SCENARIO

14.4.5.1 APPROACH 1

The process starts from first training the machine and ends with testing it for a sample set of images. The process can be divided into several modules that perform specific parts of the process and these are:

1. Dataset preparation module: The dataset that we use needs to be prepared before use. This preparation varies with the dataset used for training as different datasets have different manner of representing captions. Like the Flickr8k dataset that we use has text tokens for all the features in the images, text file that stores multiple captions of every image and the image files themselves. After this, the encoded form of the dataset is stored.
2. Caption generator module: This module is the basis for all the complete architecture of the project. This is the module where the caption is generated and model is created for the training dataset. This model is the neural network generated after the training with updated weights at all the connections. At the time of prediction, the final result is predicted by this neural network.
3. Training module: The training module basically combines the caption generator module with the prepare dataset module and the dataset itself. Here, the number of EPOCHS for which the process is to be executed is defined and the weight updation process is done repeatedly.
4. Testing module: The testing module is to test the result of the trained machine. Here, the images to be tested can be defined by the user and the results that are the predicted captions are stored in a separate file. The images can be from the dataset itself or any random image.

All these modules are performed in the above sequence and the training module is the most important module as the better the training better would be the results.

14.4.5.2 APPROACH 2

The process has overall three major modules for image caption generation process and they are the basic modules used in every image generation implementation like:

1. Dataset preparation module: The dataset that we use needs to be prepared before use. Like the Flickr8k dataset that we use has text tokens for all the features in the images, text file that stores multiple captions of every image and the image files themselves. After this the encoded form of the dataset is stored and has 6000 training images with 1000 testing and 1000 validation images. This preparation varies with the dataset used for training as different datasets have different manner of representing captions.
2. Training module: In this, the model is trained to generate the weights of the neural network that are used in the Beam Search caption generation. We have used a pre-trained model for 1 EPOCH to ease the project implementation.
3. Testing module: The images are tested here for the captions. In this, we give the location of the image and this generates the caption for the image. The accuracy of the caption is decent but can be improved.
4. Beam search module: The previously generated captions are further improved by Beam Search which checks the previous layers for the best captions. In this, we have tested the image for different index values of the Beam Search algorithm.

Finally, all these modules are combined to form an I-python notebook implementation of image caption generation with Beam Search algorithm.

14.5 IMPLEMENTATION—SCREENSHOTS

14.5.1 APPROACH 1

This involves the steps as shown in Figures 14.3–14.5.

14.5.2 APPROACH 2

This involves the steps as shown in Figures 14.6–14.11.
Next is image encoding by VGG -16 (Fig. 14.8).

1. Download the flickr8k dataset and put the text file and images in the same folder for training.
2. Create a dictionary containing all the captions of the images.

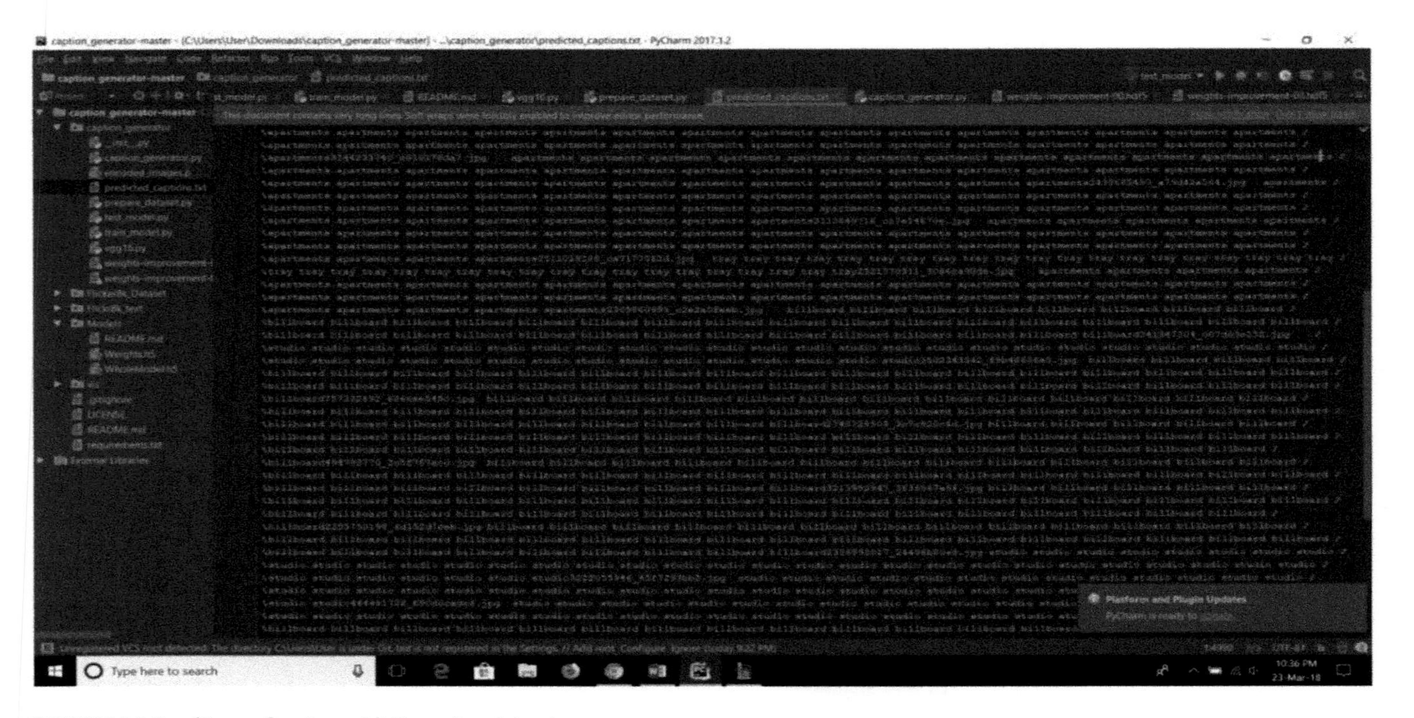

FIGURE 14.3 (See color insert.) Preparing dataset.

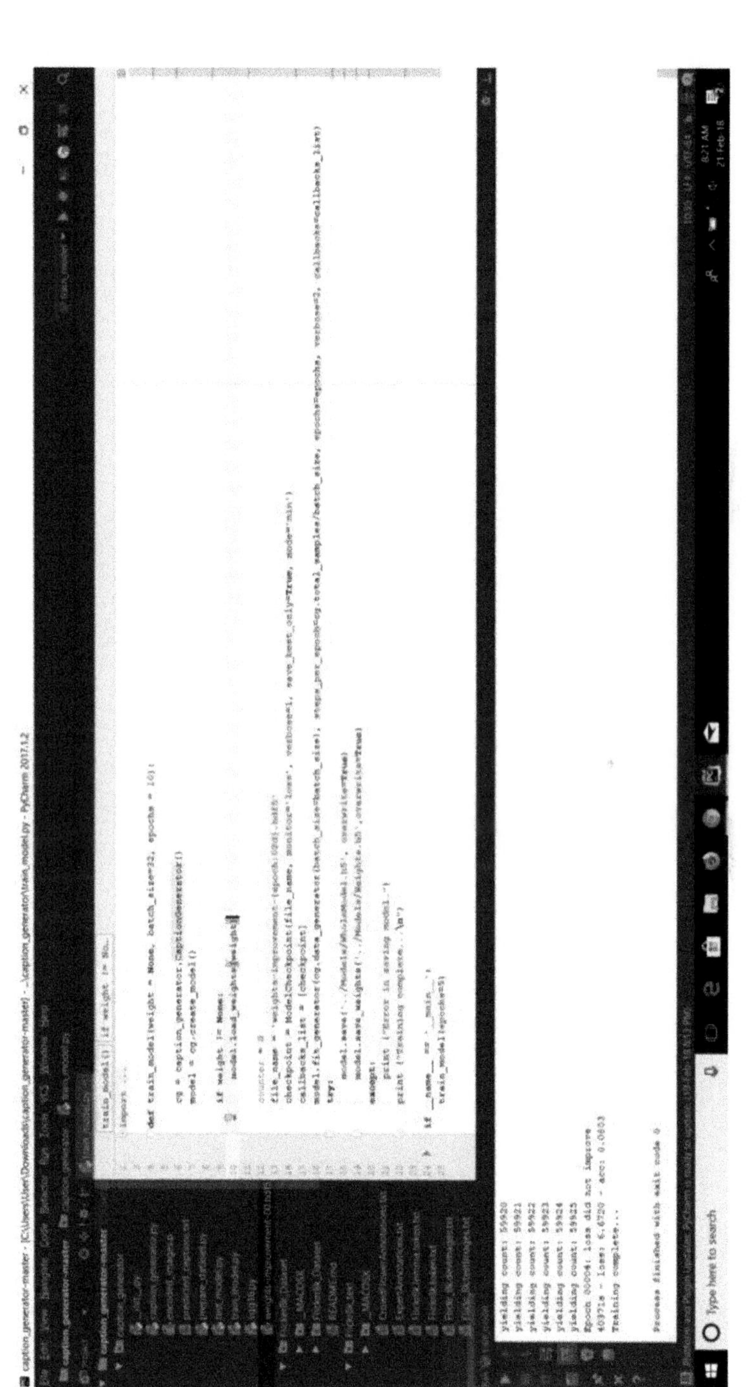

FIGURE 14.4 **(See color insert.)** Training.

FIGURE 14.5 (See color insert.) Testing.

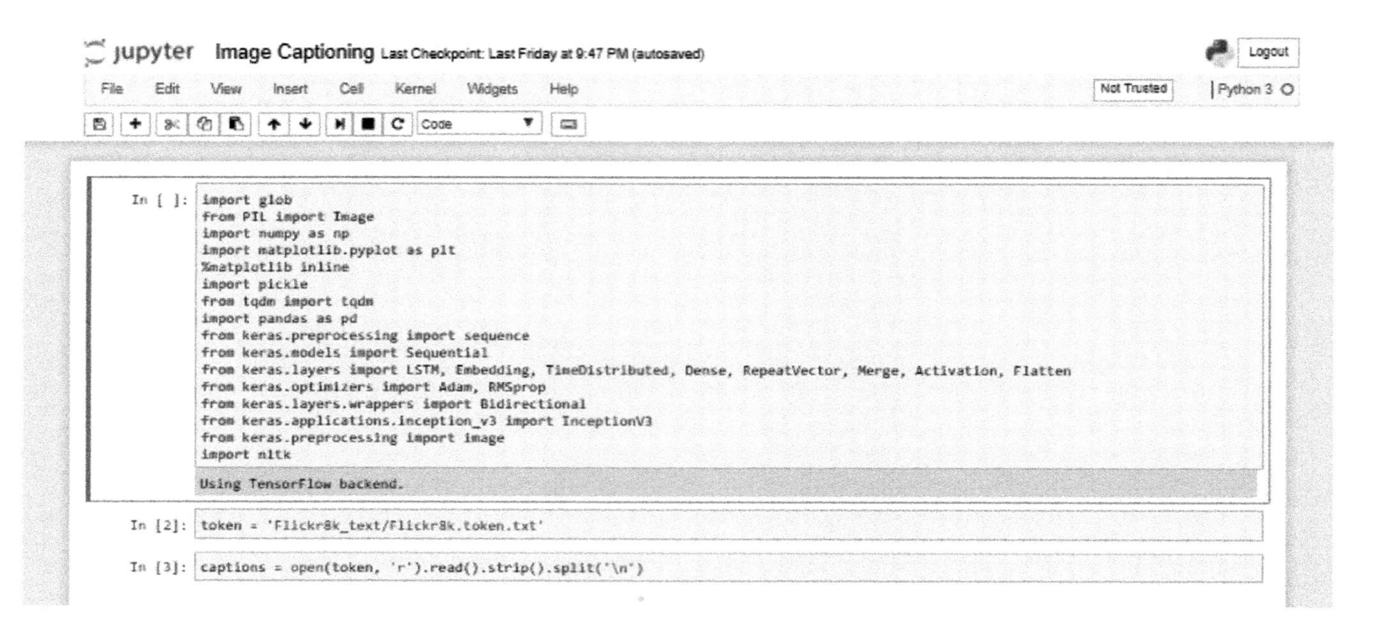

FIGURE 14.6 **(See color insert.)** Preparing dataset.

Creating a dictionary containing all the captions of the images

```
In [4]: d = {}
        for i, row in enumerate(captions):
            row = row.split('\t')
            row[0] = row[0][:len(row[0])-2]
            if row[0] in d:
                d[row[0]].append(row[1])
            else:
                d[row[0]] = [row[1]]

In [5]: d['1000268201_693b08cb0e.jpg']

Out[5]: ['A child in a pink dress is climbing up a set of stairs in an entry way .',
         'A girl going into a wooden building .',
         'A little girl climbing into a wooden playhouse .',
         'A little girl climbing the stairs to her playhouse .',
         'A little girl in a pink dress going into a wooden cabin .']

In [6]: images = 'Flickr8k_Dataset/Flicker8k_Dataset/'

In [7]: # Contains all the images
        img = glob.glob(images+'*.jpg')

In [8]: img[:5]

Out[8]: ['Flickr8k_Dataset/Flicker8k_Dataset/17273391_55cfc7d3d4.jpg',
         'Flickr8k_Dataset/Flicker8k_Dataset/2890075175_4bd32b201a.jpg',
         'Flickr8k_Dataset/Flicker8k_Dataset/3356642567_f1d92cb81b.jpg',
         'Flickr8k_Dataset/Flicker8k_Dataset/186890605_ddff5b694e.jpg',
         'Flickr8k_Dataset/Flicker8k_Dataset/2773682293_3b712e47ff.jpg']

In [9]: train_images_file = 'Flickr8k_text/Flickr_8k.trainImages.txt'

In [10]: train_images = set(open(train_images_file, 'r').read().strip().split('\n'))

In [11]: def split_data(l):
             temp = []
             for i in img:
                 if i[len(images):] in l:
                     temp.append(i)
             return temp

In [12]: # Getting the training images from all the images
         train_img = split_data(train_images)
         len(train_img)
```

FIGURE 14.7 **(See color insert.)** Creating a dictionary for captions.

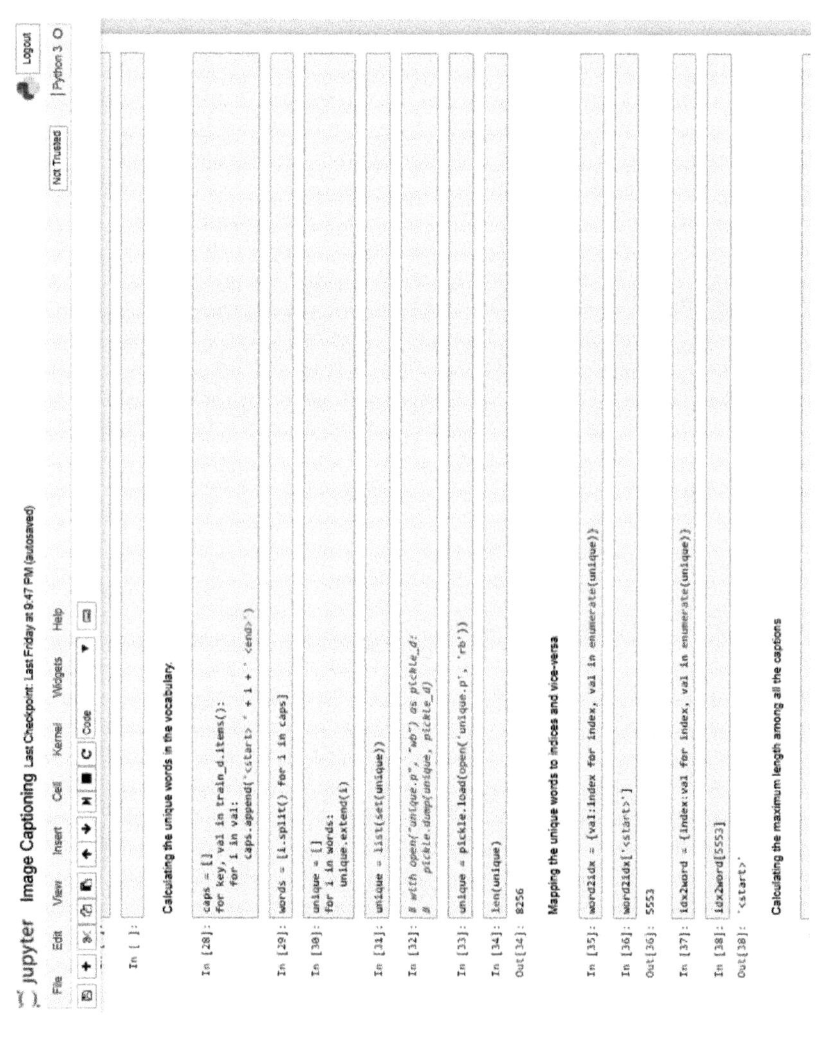

FIGURE 14.8 (See color insert.) Mapping of unique words.

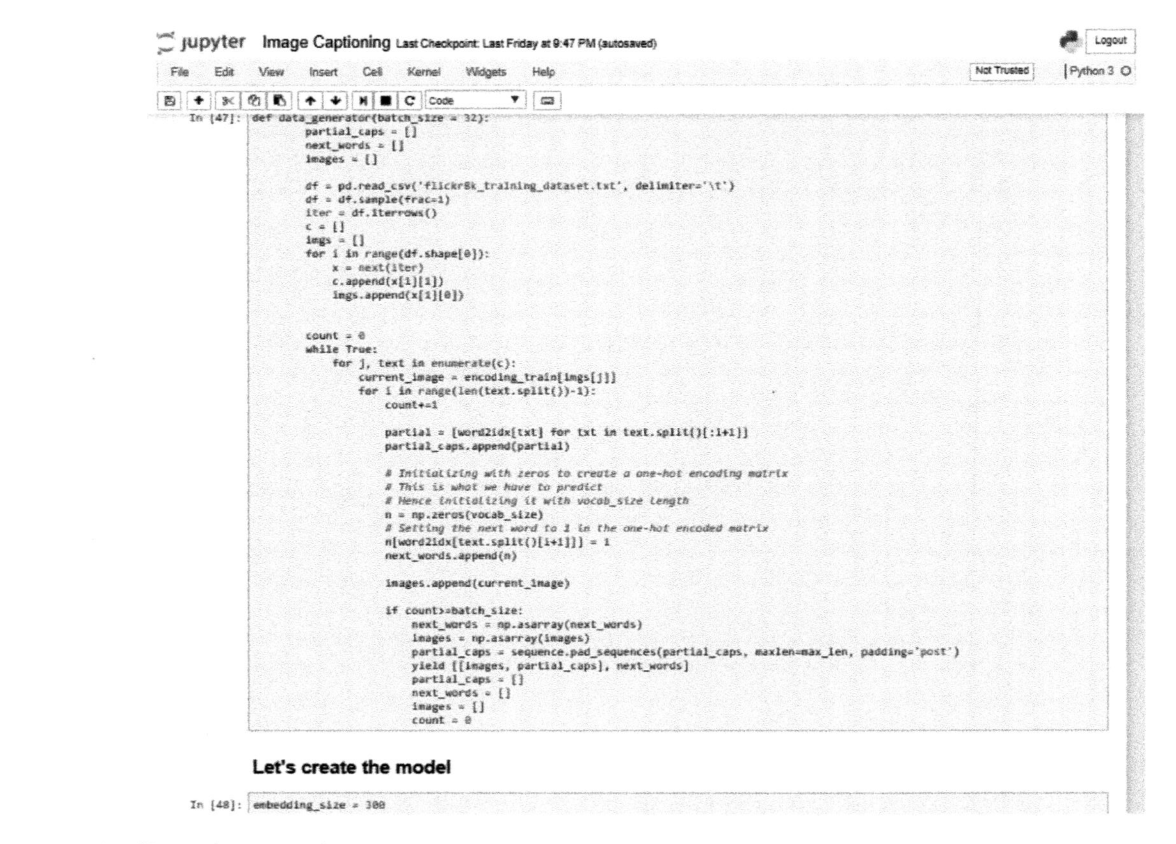

FIGURE 14.9 (See color insert.) Predicting and generating model.

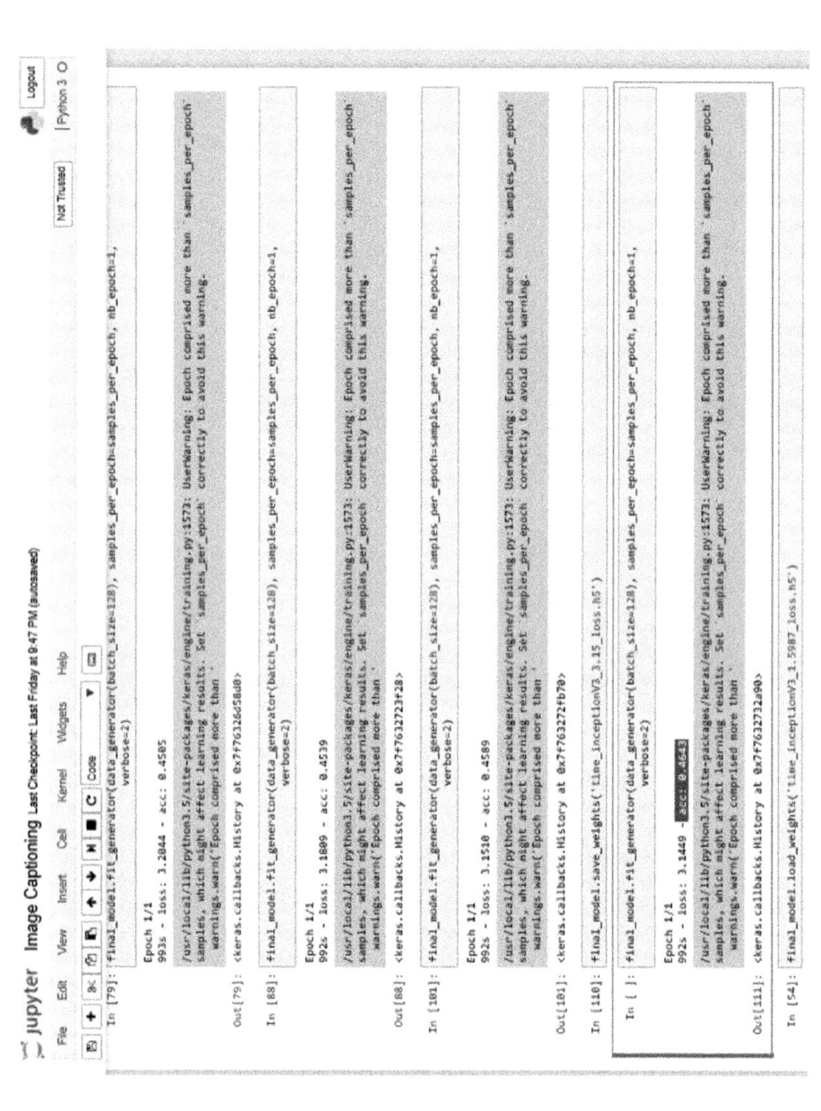

FIGURE 14.10 **(See color insert.)** Calculating loss and accuracy.

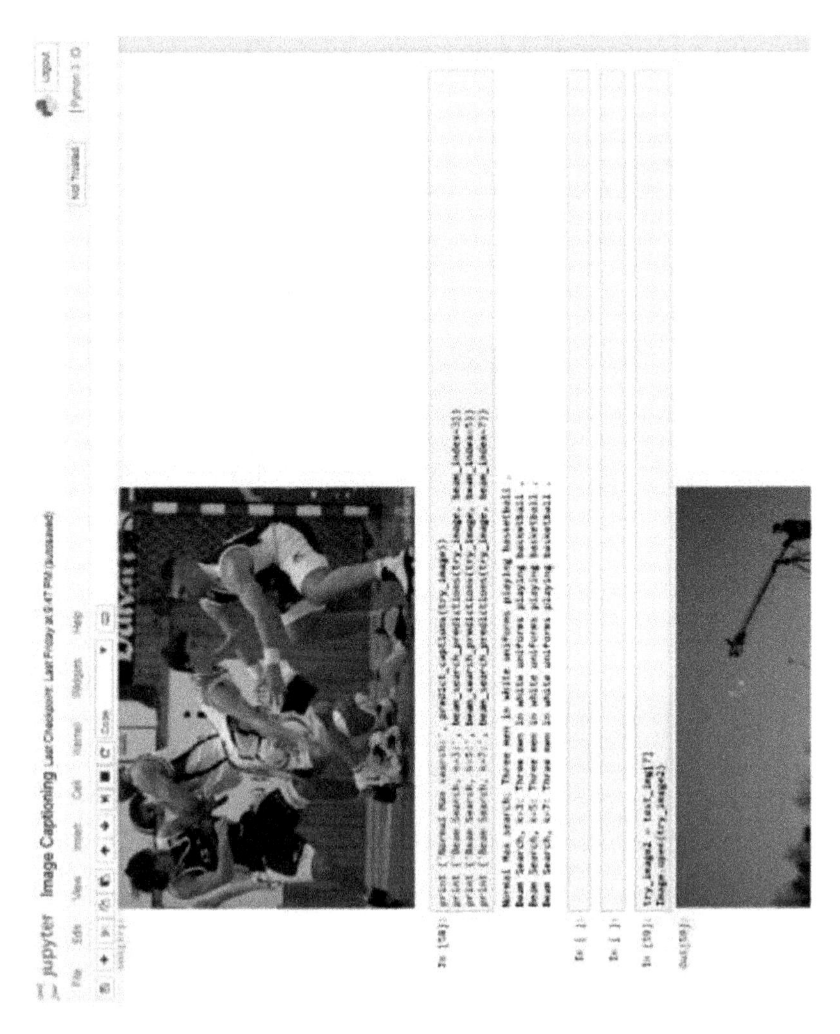

FIGURE 14.11 **(See color insert.)** Beam search for caption search results.

3. Feed the downloaded images to VGG-16 to get the encoded images.
4. Train is done using the encoded images for efficient working. The multiple captions stored in the dictionary.
5. While training uses the encoding of an image and use a start word to predict the next word. Using this formation, a model is created which helps in caption generation.
6. To generate captions testing. Now to make the generation more efficient the beam search algorithm is added in the model to search the words that are added in caption efficiently and more accurately.
7. Results are displayed on the screen.

14.6 RESULTS

After executing the program for 15 EPOCHS, the results were in accurate due to higher loss as the number of EPOCHS were not sufficient for the particular algorithm to generate the proper caption for the given image. As image caption generation is a vast process and the datasets available are too large for training through regular machines the results obtained are quite general. The accuracy results can be improved further by increasing the number of EPOCHS to at least 50 as we saw the improvement in the accuracy after one and 15 EPOCHS (Table 14.1).

TABLE 14.1 With and Without Beam Search.

Analysis	Without beam search	With beam search
Accuracy	0.0803	0.4539
Loss	6.6720	3.1809

The results obtained by both the approaches were quite different. The results varied quite a lot due to the difference in accuracy and loss of the trained model in both the approaches. The accuracy was more in the implementation of beam search algorithm and the loss was quite less even in one EPOCH while it was more in the normal implementation for four EPOCHS. The reason for this is the beam search which is quite helpful for the application of image generation as the image generation model are usually RNN and generating the best captions require the selections of best nodes by comparing the previous layers. Beam Search checks the previous layers of the neural network and then selects the best result from the previous layers which would best suit the image. The analysis shows us that beam search has significantly improved the results of the implementation. These experimental observations prove the importance of beam search in the image caption generation application.

The results obtained by both the approaches were quite different. The results varied quite a lot due to the difference in accuracy and loss of the trained model in both the approaches. The accuracy was more in the implementation of beam search algorithm and the loss was quite less even in one EPOCH while it was more in the normal implementation for four EPOCHS. The reason for this is the beam search which is quite helpful for the application of image generation as the image generation model are usually RNN and generating the best captions require the selections of best nodes by comparing the previous layers. Beam search checks the previous layers of the neural network and then selects the best result from the previous layers which would best suit the image. The analysis shows us that Beam Search has significantly improved the results of the implementation. These experimental observations prove the importance of Beam Search in the image caption generation application.

14.7 CONCLUSIONS

From the above work, we can conclude that image caption generation task can be simplified with the help of deep learning concepts of deep neural networks. The show and tell algorithm that incorporates the concept of LSTMs proves to be beneficial in simplifying the task to generate captions. Also, the conclusion can be drawn that the accuracy and the loss are directly and indirectly proportional respectively to the extent of training. Thus, it can be concluded that Beam Search can highly improve the efficiency of image caption generation as is very helpful for this application and does not increase the complexity of the project as well. As a whole Beam Search with RNN and LSTM is the one of the best implementations of Image caption generation available till date.

KEYWORDS

- **image caption generation**
- **beam search**
- **Recurrent Neural Networks**
- **Cascade Recurrent Neural Network**
- **LSTM**

REFERENCES

1. Vinyals, O.; Toshev, A.; Bengio, S.; Erhan, D. Show and Tell: A Neural Image Caption Generator. *Comput. Vis. Pattern Recogn.* **2015,** *1,* 3156–3164.
2. Sak, H.; Senior, A.; Beaufays, F. Long Short-Term Memory Recurrent Neural Network Architectures for Large Scale Acoustic Modeling. *Neural Evol. Comput.* **2014,** *1,* 338–342.
3. Chen, X.; Zitnick, C. L. In *Mind's Eye: A Recurrent Visual Representation for Image Caption Generation.* IEEE Conference on Computer Vision and Pattern Recognition (CVPR), 2015.
4. Soh, M. Learning CNN-LSTM Architectures for Image Caption Generation, 2016. https://cs224d.stanford.edu/reports/msoh.pdf.
5. Bernardi, R.; Cakici, R.; Elliott, D.; Erdem, A.; Erdem, E.; Ikizler-Cinbis, N.; Keller, F.; Muscat, A.; Plank, B. *Automatic Description Generation from Images: A Survey of Models, Datasets, and Evaluation Measures. J. Artif. Intell. Res.* **2016,** *55*, 409–442. https://www.jair.org/media/4900/live-4900-9139-jair.pdf.
6. Xi, S. M. *Weighted Feature; Region Clustering; Tmage Annotation 1.* lCCAS, 2013; pp 548–551.
7. Wu, J.; Hu, H. Cascade Recurrent Neural Network for Image Caption Generation. *Electron. Lett.* **2017,** *53* (25), 1642–1643.
8. Chowdhary, C. L.; Acharjya, D. P. Singular Value Decomposition–Principal Component Analysis-Based Object Recognition Approach. *Bio-Insp. Comput. Image Video Process.* **2018,** 323–341.
9. Chowdhary, C. L. Application of Object Recognition With Shape-Index Identification and 2D Scale Invariant Feature Transform for Key-Point Detection. *Feature Dimen. Reduc. Content-Based Image Identif.* **2018,** 218–231.
10. Chowdhary, C. L.; Muatjitjeja, K.; Jat, D. S. Three-dimensional Object Recognition Based Intelligence System for Identification, Emerging Trends in Networks and Computer Communications (ETNCC), Germany, 2011, p 76.

Index

G

H

I

K

L

M

U

V